CHEMICAL INFORMATION SYSTEMS
Beyond the Structure Diagram

ELLIS HORWOOD SERIES IN
CHEMICAL COMPUTATION, STATISTICS AND INFORMATION

(formerly The Ellis Horwood Series in Chemical Information)

CHEMICAL INFORMATION SYSTEMS
J. ASH and J. E. HYDE

COMMUNICATION, STORAGE AND RETRIEVAL OF CHEMICAL INFORMATION
J. E. ASH, P. A. CHUBB, S. E. WARD, S. M. WELFORD and P. WILLETT

CHEMICAL INFORMATION SYSTEMS: Beyond the Structure Diagram
D. BAWDEN and E. MITCHELL

CHEMOMETRICS: Applications of Mathematics and Statistics to Laboratory Systems
R. G. BRERETON

COMPUTATIONAL METHODS FOR CHEMISTS
A. F. CARLEY and P. H. MORGAN

NAMING ORGANIC COMPOUNDS
E. GODLY

CHEMICAL NOMENCLATURE USAGE
R. LEES and A. F. SMITH

DESIGN, CONSTRUCTION AND REFURBISHMENT OF LABORATORIES
R. LEES and A. F. SMITH

HANDBOOK OF LABORATORY WASTE DISPOSAL
M. J. PITT and E. PITT

COMPUTER AIDS TO CHEMISTRY
G. VERNIN and M. CHANON

COMPREHENSIVE DICTIONARIES OF CHEMICAL SCIENCE:
Volume 1: Physical Chemistry
L. ULICKY and T. J. KEMP

CHEMICAL INFORMATION SYSTEMS
Beyond the Structure Diagram

Editors

DAVID BAWDEN Ph.D.
Principal Information Scientist
Pfizer Central Research, Sandwich, Kent

ELEANOR MITCHELL Ph.D.
Scientific Officer
Cambridge Crystallographic Data Centre, Cambridge

ELLIS HORWOOD

NEW YORK LONDON TORONTO SYDNEY TOKYO SINGAPORE

First published in 1990 by
ELLIS HORWOOD LIMITED
Market Cross House, Cooper Street,
Chichester, West Sussex, PO19 1EB, England

A division of
Simon & Schuster International Group
A Paramount Communications Company

Printed and bound in Great Britain
by Bookcraft (Bath) Limited, Midsomer Norton, Avon

British Library Cataloguing in Publication Data

Chemical information systems: beyond the structure diagram. —
(Ellis Horwood series in chemical computation, statistics and information).
1. Chemistry. Information retrieval
I. Bawden, David. II. Mitchell, Eleanor
025.0654
ISBN 0–13–126582–2

Library of Congress Cataloging-in-Publication Data available

Table of contents

Introduction: Beyond the structure diagram

David Bawden, Pfizer Central Research, Sandwich, Kent, England.
Eleanor Mitchell, Cambridge Crystallographic Data Centre, Cambridge, England.

INTRODUCTION TO THE CONCEPT

Going 'beyond the structure diagram' necessarily implies the concept of *integration*. Integration of the 'traditional' functions of chemical information systems (structure match, substructure search, compound registration, etc.) with the functions more usually associated with 'computational chemistry': molecular orbital and molecular mechanics calculations, molecular graphics, property prediction, correlation analysis, and so on. This is a point which will be made throughout the chapters in this book.

No true integration is possible without the availability of a flexible and reliable system for storing and searching a complete and unambiguous chemical structure representation. Once this is achieved the door to integration 'beyond the structure diagram' is open.

Carlos Bowman made this point clearly in 1975:

> Once the step has been taken to fully represent the chemical structure in a computer-based data file, we have acquired the flexibility and capability to do many things previously unimagined . . . It is now possible to relate structure and other information, such as physicochemical properties and much other information about a compound. [1]

The development of chemical information systems has been almost exclusively a story of the development of special purpose systems or routines, followed by attempts to integrate them at a later stage. Thus during the 1970s there was a sustained development of systems for such purposes as:

structure searching,
reaction retrieval,
synthesis prediction,

molecular modelling,

conformational analysis,

but all developed essentially independently.

Following the introduction of computerized chemical information systems into many organizations during the 1980s, a large number of examples of structure systems with other functionality began to appear. Most of these were concerned with integration of structure searching with the retrieval of associated biological or physicochemical data. Similarly, it was only in the late 1980s that database access became a major concern for molecular modelling systems, despite the proven value of structural databases such as the Cambridge files. Again, in the areas relating to computer-assisted synthesis planning, it was only in the late 1980s that some elements of overlap began to appear between systems for synthetic analysis, reaction retrieval, and substructure search.

It is remarkable, as we have said, how quickly a concern for integration of information arose, following the widespread adoption of efficient substructure searching systems during the early 1980s. Of the papers given at the major multipartite chemical information conference in 1987 [2], more than half dealt, in some respect, with the theme of integration, with major contributions from Williams and Franklin and from Hagadone. In particular, many of these echoed Bowman's views, by describing moves towards integration of chemical structure with physical and biological data. Papers of this sort came from Akzo (de Jong and Diebel), Ciba-Geigy (Shlevin), Glaxo (Allen), Fisons (Magrill), and Pfizer (Bawden *et al.*), and in the public database area from Beilstein (Jochum) and ECDIN (Norager). Other aspects of integration included the linking of structures with text documents (Shlevin), integrated chemical reaction information (Grethe *et al.*, Sieber, and Vleduts and Gould), prediction of property (Gasteiger), structure–activity correlation (Meyer), and links between structures and spectral data (Bremser).

To some extent, the chapters in this book form a continuation and extension of this theme. That conference, however, had as its theme 'Chemical Structures. The International Language of Chemistry', emphasizing the centrality of the structure diagram representation. The 1989 conference that led to this book, in seeking to go 'Beyond the Structure Diagram', aimed to look beyond what was possible for present-day integrated systems, to future prospects.

It is important to note, however, that we are going *beyond* the structure diagram, but *not* away from it. Many factors were at work in promoting integrated systems for chemical information during the 1980s. These include improvements in computer technology, and specifically of system software tools, and the increased sophistication and expectations of the user community. Nonetheless, it is important to recognize the centrality of computer-handling of structure diagram representations in providing the basis for integrated systems.

So, the phrase 'beyond the structure diagram' expresses the integration between chemical structure searching systems and other chemical information systems (in the broadest sense). Chemistry is a uniquely information conscious subject area, and the central role of the structure diagram representation is largely responsible for this. Nonetheless, there may be some lessons to be learnt in the optimal use of information technology for information handling in other areas.

In practical terms, going 'beyond the structure diagram' within chemical information systems has two mutual benefits for two distinct groups of systems and their user groups. It gives users of structure searching systems convenient access to a much wider range of chemical computation facilities. At the same time, it improves and broadens the scope of the other computational chemistry functions, by providing improved interfaces and database facilities, and by making them accessible to a wider user community.

INTRODUCTION TO THE BOOK

The chapters in this book exemplify the two benefits, mentioned above, in going beyond the structure diagram: access for information system users to a wider range of computational facilities, and greater accessibility for advanced chemical computing systems.

Ward gives an overview of the impact of information technology in general, and chemical information systems in particular, on pharmaceutical research. She exemplifies this by developments at Glaxo Group Research, emphasizing the concept of the scientific workstation for providing convenient user access across a spectrum of information services.

The introduction of techniques for searching for three-dimensional chemical structures, whether in the form of stereochemically specified structures or as databases of atomic co-ordinates, is described in three chapters.

Barnard, Cook and Rohde review the way in which stereochemical information is encoded, stored and searched within 'conventional' substructure searching systems. There are a number of approaches, at varying levels of sophistication. It is to be hoped that adoption of the Standard Molecular Data (SMD) exchange format, which allows for several alternative stereochemical representations, will enable more efficient transfer of information than is possible at present.

Moock, Christie and Henry exemplify the trends towards 3D searching in commercially available systems, by a description of the MACCS-3D program. This illustrates clearly the advantages of linking 2D substructure searching with 3D conformational retrieval.

Grindley *et al.* discuss a variety of techniques for graph matching in files of 3D structures, applied to the small molecule structures in the Cambridge Structural Database, and to macromolecules in the Protein Data Bank. These techniques are directly analogous to those used in 2D substructure searching.

Mitchell, Allen and Kennard review the operations of the Cambridge Structural Database System, an increasingly important tool for both fundamental and applied research. They emphasize the integration of bibliographic, 2D structural and 3D structural information, and the value of systematic analysis in a carefully validated set of data.

Reactions and synthesis planning, and their integration with structure systems are discussed in two contributions.

Hopkinson, Cook and Buchan review reaction retrieval systems, with particular reference to ORAC. They point out the degree of commonality between systems for retrieval of reactions, and those for retrieving single structures, and illustrate some of the particular problems of reaction searching, such as identification of tautomers and handling of multi-step reactions.

Loftus gives an overview of software for assisting the planning of organic syntheses, and describes ICI's experience with EROS and with LHASA, two very different kinds of synthetic planning system, and with CHIRON, a program for synthetic analysis. Although he comments that progress towards the ultimately desirable system is frustratingly slow, this paper makes clear the potential value of such systems.

The important area of property prediction from structural considerations is covered, from different viewpoints, in three chapters.

Ball *et al.* describe a very advanced computational chemistry system for molecular modelling and property prediction. Their 'polymorphic programming environment' includes aspects of operating system functionality, programming language, expert system, and molecular simulation.

Rouvray describes the use of topological indices for property prediction. His wide-ranging overview sets the use of these mathematical descriptors into context, and shows both their theoretic basis and practical value.

Johnson reviews the whole area of molecular similarity, and its applications in predicting chemical and biological properties. In particular, he shows similarity-based property prediction to be an important example of a molecular similarity space, and uses this mathematical concept to set particular approaches into context.

Finally progress towards integrated chemical information systems, and the likely future shape of such systems, is reviewed by Bawden, and exemplified by developments at Pfizer Central Research. He emphasizes the continuing central role of the structure diagram representation in future chemical information systems. This will be incorporated within systems allowing both sophisticated retrieval facilities, far beyond present-day substructure search capabilities, and also links to routines for modelling, calculation, prediction, planning and correlation.

ACKNOWLEDGEMENTS

We are grateful to our colleagues on the organizing committee (Judith Dalton, Steve Hull and Lesley Tewnion) for their contributions in ensuring the success of the CSA 1989 Annual Conference, of which this book forms the permanent record.

REFERENCES

[1] C.M. Bowman, 'The development of chemical information systems', in 'Chemical Information Systems', J.E. Ash and E. Hyde (eds.), Chichester, Ellis Horwood, 1975.

[2] W.A. Warr (ed.), 'Chemical Structures. The International Language of Chemistry', Berlin, Springer-Verlag, 1988.

Overview

Chemical information handling: a management view

Sandra Ward, Glaxo Group Research Information Systems and Services Division.

THE CSA

I was delighted to be given the opportunity to participate in a CSA Conference again, albeit briefly. I seem to remember way, way back, as a very new Information Scientist – and of course a very young one – attending an early meeting, in Loughborough, which I think was an AGM of the Chemical Notation Association (CNA) (UK). This group was, of course, the UK branch of an American Association, set up to promulgate the use of WLN (Wiswesser Line Notation) for the computerized handling of chemical structures. It brought together, in an extremely creative group, the UK's chemical and pharmaceutical companies, largely through the vehicle of their information departments. Members of the CNA not only developed training courses in WLN and pioneered the creation of the Fine Chemicals Directory as a WLN-based index to commercial available organic chemicals index. They also, from the start, and throughout the transformation of the CNA into the CSA in 1982–1983, have run highly successful and thought-provoking conferences. These have not only kept people up-to-date with the developments in computerized chemical structure handling, particularly the emergence of graphical techniques, but have included how chemical structure handling software could be creatively employed in organizations both within company information systems and for accessing published information. The integration of databanks of structures with biological and physicochemical property information systems to provide systematic and adaptable records of research has also been heavily discussed. The creative use of such data for structure–property analysis and prediction has continued to be a CSA preoccupation – as has reaction indexing, computer-assisted chemical synthesis software and the development of the large synthesis databases that are now proving of such benefit to the organic and development chemist. The CSA has always been 'beyond the structure diagram' . . .

Not only has the CSA wanted to ensure that its conferences have concentrated on the integration of chemical structural information handling with other related data handling activities, it has also tried to integrate those professional groups most interested in the area – information scientists involved in both internal and published information provision, practising chemists and biologists, computational chemists and data processing professionals. The CSA has always been conscious that effective and integrated information systems will only be realized if the interests and objectives of all these groups are debated and if the skills of these professionals can be jointly utilized. Early conferences concentrated on giving the chemical information scientist an understanding of the data processing role in systems analysis and design. More recent ones have moved beyond producing a common vocabulary to considering information handling developments from the perspectives of all these professional groups. It is particularly welcome to see computational chemists and information scientists increasingly converging on the conference circuit.

The themes of the book are how we are now moving to an era in which storage and retrieval utilizing the two-dimensional structural representation is complemented by techniques for searching for particular stereospecific configurations and by techniques for three-dimensional substructure searching utilizing databases of 3-D coordinate data. Other chapters will look at the utilization of structural data – for modelling and for calculation of molecular properties and for reaction planning. Finally the goal of the integrated system and its feasibility will be explored.

As an Information Manager within the pharmaceutical industry, and one who spent much time in the past coding compounds into WLN, I am still enormously interested in the development of information systems for the research chemist and biologist. I do not, however, intend to give what would in my case now need to be a layman's view of the state of the chemical information art. What I am now most involved with and what I intend to cover in this chapter is the pressures on the pharmaceutical industry, the potential of IT to support the industry's research and development function, and how chemical information handling must be considered with other IT needs for its investment priority. I will be writing from the perspective of an information systems planning project with which I have recently been involved.

I want to do this because – although companies like Glaxo are major investors and therefore major influencers of the chemical information scene – new developments of the type covered in this book must be seen in the context of a company's overall information needs, must be justified in business terms and cannot simply be adopted because they are there.

THE PHARMACEUTICAL INDUSTRY

The pharmaceutical industry is a high risk, high investment, high technology industry – and highly competitive. The industry and its individual companies are also highly international.

The financial analysis of the industry worldwide makes fascinating reading. For 1988 detailed figures are only just emerging but these show the total market in 1988 to be £65 000 million (Table 1).

Worldwide, *ca*. 28% of this market is in the USA, 24% in Japan and about 27% in Europe. When analysed by company, we see that the leading company, Merck, has only a 4.1% share of the world market. A number of companies, including my own, hold between 2 and 3% of the market.

Table 1 – Market share (%) – pharmaceutical companies 1988

Merck	4.1
Glaxo	2.9
Ciba-Geigy	2.9
HOECHST	2.7
American Home Products	2.7
Smith Kline Beckman	2.3
Pfizer	2.2
Sandoz	2.1
Bayer	2.1
Lilly	2.0

The industry is thus highly fragmented – 30 companies have shares greater than 1% in contrast to, for example, the car industry where the top five car manufacturers account for over 90% of world sales.

To quote Ray Vagelos, Merck's current Chairman [1] . . . 'To be No. 1 and so little is ridiculous – if we could double our market share then we would have the revenue stream that would allow us to work in all the interesting fields at once. At the moment we have to race to be ahead of the competition.'

The money which a novel, safe and effective medicine can earn for a company can be substantial. Zantac sold over £1000 million in 1988. And, a major new drug can cause rapid changes in industry leadership.

But 'blockbuster' drugs are rare:

- Only 52 new chemical entities (NCEs) were introduced into their first markets in 1988 (Table 2) – the majority from different companies. Increased regulatory requirements and development costs are probably the principal constraints on the industry's record of product introduction. But, the increasing complexity of research targets and the levels at which we want drugs to exert their action within the body must also be a factor [2] ;
- For every drug which enters the marketplace, a company may have synthesized 10 000 which do not [3] ;
- The development phase of an NCE from initial discovery to marketing now often extends over more than 12 years, discounting the work which precedes the initial discovery. This is made up of:

  – establishing activity and pre-clinical safety 2–10 years, and
  – undertaking clinical safety and efficacy testing 3–9 years, depending on whether the drug is an acute or chronic therapy;

- The costs of developing a new drug are estimated as being in the order of £100 million to bring a drug into full production and worldwide marketing (£1 million in research, £10 million in clinical development);

Table 2 – New chemical entity introduction

Early 1960s	90–100 p.a. •
Early 1970s	70–90 p.a.
1984	43 *
1985	57
1986	47
1987	58
1988	52

• *Trend and Changes in Drug Research and Development*
* *Scrip*, 2nd April, 1989, review p. 14

- Companies are now spending 10–15% of turnover on research and development – a proportion which is growing and which, of course, is much higher than other industries. The industry's R&D costs are increasing at a rate of 20% per annum in a market which is growing at approximately half this pace. These increased costs reflect *inter alia* an increase in manpower employed in R&D. In 1982, the top 10 pharmaceutical companies (ranked by R&D manpower) employed 26 145 staff. By 1987, this number had increased by 40% to 36 300;
- Only a relatively small proportion of these R&D costs involve pure drug invention – the majority (*ca.* 70–80%) goes into the development process, made up roughly as:

Synthesis	12%	Biological testing	20%	Regulatory	3%
Toxicology	10%	Bioavailability/ Pharmacokinetics	4%	Miscellaneous	10%
Chemical development	8%	HVS	3%		
Clinical work	17%	Pharmacy	12%		

  The miscellaneous category probably includes chemical information work! This reflects, of course, the increased requirements of regulatory authorities with extensive safety testing needs and newer requirements such as pharmacokinetic and efficacy benefit studies, and a greater emphasis on new drugs for chronic rather than acute conditions, requiring vastly more data;
- Over the last 10 years, a clear trend towards more focussed R&D effort has emerged among the leading pharmaceutical companies [3]. Whereas in the 1970s companies would commonly have R&D programmes in 10 or even 20 areas, today most companies have reduced the areas under research to about a half dozen or so;
- But, no matter how much a company throws at research, it cannot guarantee that it will come up with another bestseller. Contrast Britain, which has launched 10 of the 50 top selling drugs, with France, which spends as much on R&D but has not one drug in the top 50 [2];
- The industry's profits, prices, and promotional costs are heavily controlled by government.

Few industries have the same high risk profile as pharamceuticals, yet the challenges are still there. Even now, perhaps only 50% of therapeutic needs can be met with existing drugs. So R&D in the pharmaceutical industry will continue to be expensive and high risk, with a rare occurrence of a very high reward.

To improve their success, pharmaceutical companies must therefore, as far as R&D is concerned, look for more effective approaches to research and must aim to streamline their high cost development processes. In doing so, companies have obviously looked to the potential of information technology to facilitate drug design and development.

The proportion of pharmaceutical industry R&D expenditure directed towards information technology investment is not well-documented. It seems probable, however, that a figure of *ca.* 10% is not unreasonable. Thus our £65 000 million world market equates (very roughly) to £650 million pounds per annum for provision of IT based services! And this only for the R&D component of the industry. No wonder there are an increasing number of companies developing software and hardware in support of drug research and development. And no wonder that the pharmaceutical industry competes heavily for well-qualified and imaginative IT staff.

IT IN PRODUCT DEVELOPMENT

The development activities of the industry involve large teams and many subject disciplines (from development chemistry through pharmacy to toxicology and medicine), and may occur across a number of international locations. All activities must be carefully planned and co-ordinated. Investment in IT has up to now concentrated on making this process as efficient as possible. IT has been used to streamline the large scale routine and repetitive processes associated with high volume testing and data analysis. Laboratory information management systems, process control systems, project management systems are all featured within the industry's IT portfolio. An important new development in applications software is integrated technical information management systems [4]. These aim to link the laboratory data obtained from laboratory information management systems and other research to the data analysis and reporting required 'en route' to the regulatory submission. The facility to deal with compound documents, containing text, data, chemical and other graphics, in an efficient way and via completely electronic techniques will provide companies who can exploit their investments in this area effectively with real competitive edge. It is also conceivable that a stronger theoretical and modelling approach to drug development may eventually make it feasible to dismantle some of the costly processes of safety and efficacy testing.

IT IN RESEARCH

In looking for more effective routes to research success, information technology investment is again proving of critical importance to the industry and has been focussed on a number of areas.

The traditional 'spray and pray' or 'molecular roulette' approach has required considerable resource to establish corporate databases of structures synthesized or acquired for testing, complementary databases of biological and physicochemical data to support the review of structure–activity trends and patterns, and laboratory automation systems to improve throughput of analytical and biological tests.

Most companies have also invested in access to reaction indexing systems either for published information retrieval or for storage of in-house laboratory records. A more selective number of companies have developed or are exploiting software for computer-assisted chemical synthesis.

The identification of chemical structures can be facilitated by the online availability of reference data and via expert systems. It is, however, in the area of computer-aided drug deisgn that companies hope to realize the greatest gain. The tremendous progress made in protein purification methods, amino acid sequencing, nuclear magnetic resonance protein crystallography, and computer graphics has, and is, producing accurate and three-dimensional information on receptor-sites. Molecular modelling, conformational analysis, and computer graphics allow the production of models for a wide range of chemical structures, the prediction of molecular properties, and the examination of the interaction of a variety of compounds with the receptor site of interest. Utilization of these techniques to determine computationally the structures of compounds that are likely to be selective and effective medicines (and also which compounds should not be made) can provide companies who can exploit these techniques effectively with increased probability of success in drug invertion. Hopefully this will be possible at a smaller human cost and a shorter lapse time. Although the costs of the technology itself are significant, developments in workstation technology including the advent of graphics supercomputers (Apollo, Ardent Titan, and Stellar) are already covering the costs of investing in near-Cray computer power combined with ultra-high-speed three-dimensional graphics.

IT thus has potential benefits throughout research and development. But IT benefits are not realized overnight. Large projects will take many man years of effort to bring to successful completion and the costs of these types of investments must relate to the business benefit that they confer. Glaxo Group Research Ltd., like most other companies, must carefully prioritize its IT investment opportunities to match its key business objectives. The cost of introducing systems such as CROSSBOW is far less than for systems such as MACCS, and hardware–software investment in areas of interest to chemists continues to be a costly exercise.

GGR'S INVESTMENT IN CHEMICAL STRUCTURE SYSTEMS

In the past few years, Glaxo Group Research Ltd. has made investments in research computing which are broadly similar to those of other pharmaceutical companies.

- MACCS software used for searching an in-house registry file, the Fine Chemicals Directory, Stores compounds, and Medchem;
- A commitment to support MDL's MACCS-3D project;
- An investment in REACCS and the MDL reaction indexing databases – JSM, Organic Synthesis, Theilheimer and MDL's Current Literature File, ISI's Current Chemical Reactions database and CHIRAS (MDL's database development for access to stereo-specific synthetic routes);
- Database systems for handling information on molecular properties of in-house compounds, from basic physicochemical properties (melting point, boiling point solubility) through to biological data in summarized form;
- Investments in chemometrics, molecular modelling, and protein crystallography; utilizing a mixture of commercial and in-house software (including in-house

availability of the Cambridge Crystallographic file and Brookhaven sequence databases);

- Sample management systems for managing in-house research samples, both storage and co-ordination of biological testing;
- Access to commercial published databases such as CAS-Online, both by information scientists and bench chemists.

GGR's investments are broadly similar to those of other pharmaceutical companies, with in-house substructure access to 2-D structural information with conformational display and access to related data.

Gaps in our portfolio include:

- in-house reaction databases for chemical development and research
- laboratory information management systems for analytical chemistry and chemical development
- 3-D databases of chemical structures
- Markhush output from structure and data searching
- online facilities for requesting spectral measurements, compound registration and compound samples
- improved facilities for access to physicochemical data
- automated techniques for compound screening equipment interfaces, and data capture for biological testing involving robotics, barcoding, and the automatic compilation of reports
- generic systems for the storage and processing of raw data generated from screening operations linked to automatic data capture and testing
- consistent approaches to the management, storage and reporting of data from research projects linking to summary level biological systems. These are under development.

As GGR considers the gaps it has perceived together with new developments in chemical information, it must select carefully new areas for investment which it feels will give the company real competitive advantage.

A decision to convert from a Prime to a VAX hardware base in 1986 had given us the twin problems of an immediate backlog in software conversion and a focus of high interest in the opportunities created by moving to Digital-based computing.

We therefore in 1988–1989 have taken the opportunity

(a) to review needs for IT across the business
(b) to undertake a particular analysis of document production in order to identify where this could be facilitated by improved investments in technology
(c) to produce an IT investment plan for the next 12 years or so, together with management mechanisms which would support the achievement of this plan and its review and extension in subsequent years.

GGR'S ARCHITECTURE PROJECT

This review, officially entitled GGR's Applications and Document Architecture project, has just finished.

The project had a number of substantial inputs:

- application plans outlining *systems development goals for discrete business areas*;
- *user workshops* covering all the company's principal system interests. These brought together users from different parts of the business to encourage cross-fertilization of ideas and to consider especially needs for corporate information systems;
- the production of high level *information flow models* and data analyses;
- the *analysis* of the production of a *regulatory submission* from data origin (in machine or human recorded form) to the appearance of the information on the submission page. This covered data and document flows, document volumes and production methods.

The resulting opportunities were assessed then for priority against a decision model developed from the business objectives and critical success factors of the company and those priority projects which, given present resource levels, could be initiated in the next three years or so were featured in the GGR information systems corporate development plan.

Commercial sensitivity precludes a detailed presentation of the plan here but overall themes are relevant.

The overall breakdown of projects into business areas was as follows:

Research	20%	Engineering	3%
Development	42%	Corporate affairs	10%
Medical	13%	Management services	6%
Company-wide investments	6%		

The priority which GGR will need to give to product development including clinical data processing reflects the significant number of compounds currently in development and the need to ensure the streamlining of all development activity via integrated functional information and sample management systems.

As well as the applications priorities other areas identified as key were:

- IT education and training;
- integration;
- document production;
- strategies for the relative roles of pc and mainframe software;
- simpler user interfaces;
- strategies for central versus user responsibilities for IT related activities.

Discussion and decisions in these areas included both research and all other company activities.

Integration emerged as an important user need and the following priorities have been identified:

- data processing and document production – facilities to move easily from data analysis to incorporation of data into documents and facilities for moving both easily – throughout the company and the Glaxo group;

- integrated end-user computing tools which interface with all levels of computer systems to facilitate analysis, presentation and exchange of data;
- 'vertical' integration between the different data processing levels within a directorate, ensuring that systems for generating and handling different data levels within a given functional area 'fit' together in a well-structured and seamless way;
- 'horizontal' integration between systems in different functional areas – the linkages between systems handling data on development compounds are of most importance;
- the redefinition or creation of corporate information systems, which hold information needed by several business functions or which should hold data which is currently duplicated in a number of systems, so that these corporate systems occupy the most effective position in our overall information systems architecture.

Within research, priority for IT investment will be given to further building blocks in an integrated approach to research information handling. Thus we will aim to provide a totally structured approach to biological data computerization, also to provide general utilities to ensure that all computerized systems requiring access to the company's core chemical systems can do so, and to introduce personal computer software for local processing and display. Our corporate priority is therefore to provide links which enable us to maximize the usefulness of already available machine readable data.

THE SCIENTIFIC WORKSTATION

In addition, there is one major project being undertaken by our American colleagues which we intend to consider seriously – the Scientific Workstation (SWS) project. This again supports the integration theme. SWS is a joint development initiated by MDL, Digital, and Glaxo Inc. Running on Digital's VAX workstations, *inter alia*, the system will provide an integrated working environment for the research scientist through the provision of a single-user menu-driven interface which can act as a front end to company and external computing resources (Fig. 1). The product is being developed under DEC windows by MDL.

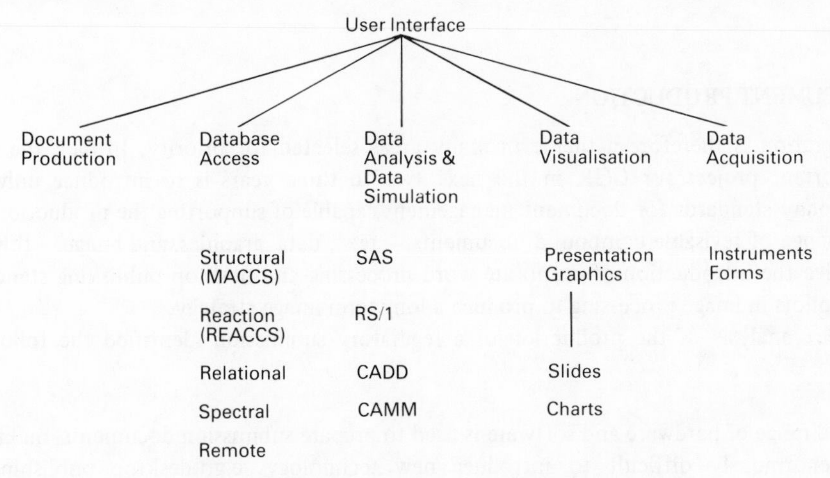

Fig. 1 – Scientific work station functionality.

The prospect of the chemist's workbench is exciting. Key to the system is the user interface. This contains terminal capabilities for graphics-based applications such as MACCS, REACCS and modelling software. It will run presentation graphics such as RS/1 and Tell-a-graf, and forms and applications drivers for programs such as ORACLE.

The functionality contained in the interface will be extensive – a client from the workstation will be able, via a form, to generate a chemical and/or database query. The system will arrange for the search to be executed locally, on the mainframe or externally. The results will be returned to the workstation where the user has the option of printing, including the results in a report, mailing this to colleagues, or undertaking additional analysis of this data using data analysis or modelling tools on the mainframe.

The benefits to the research scientist of using a single consistent user interface for accessing all application programs is obvious – as are the related benefits of reduced training requirements. Systems efficiency gains from operating in a distributed environment are also anticipated. Because the project is based on Digital's architectures including their compound Document Architecture it should be of wide interest to companies whose systems are based on DEC equipment. DEC, for example, and, one anticipates, other third party software houses will make other applications software available in this environment, including projects to produce compound documents.

A generalized drawing editor will provide

- connection table based graphics;
- sketching-based graphics with no chemical significance;
- forms design tools and links to enhance data structures;
- support for mathematical formula and multiple font styles.

Although the product has been developed with the chemist in mind, there is no reason why its use cannot be extended to other areas for (a) user interface (b) pulling information together from a variety of sources, but the costs are likely to be high and introduction of a product such as SWS into routine use will require extremely sound justification.

DOCUMENT PRODUCTION

Integration is therefore a theme in many areas selected for priority. In fact, the most important project for GGR in the next two to three years is to introduce universal company standards for document management capable of supporting the production and exchange of revisable compound documents – text, data, graphics, and image – this will involve the introduction of corporate word processing and desktop publishing standards and pilots in image processing to produce a long term image strategy.

Our analysis of the production of a regulatory submission identified the following issues:

- a range of hardware and software is used to prepare submission documents, making it enormously difficult to introduce new technology, e.g. desktop publishing or document image processing;

- data re-entry and cut and paste are still considerable features of the document production life cycle, particularly as data from different systems is analysed and reported;
- the merging of text, data, and graphics is extremely difficult because of the volumes involved and the widespread use of different items of hardware and software;
- the professional/para-professional interface is ineffective;
- inconsistencies in the format and content of different documents necessitate re-writing, reformatting and additional editing;
- international exchange of documentation is not yet practicable for electronic revisable compound documents.

These issues are not, I am sure, ones which Glaxo alone faces and they are certainly ones which are relevant to the CSA.

The ability to manage document production, storage, distribution, filing, and retrieval of documents is likely to be a key factor in bringing compounds in full development to registration in future. Developments in systems which will handle compound documents and the emergence of ISO standards such as ODA and ODIE which provide the basis for the interchange of revisable compound documents between different computer systems are making the time ripe for reviewing activity in this area. It will become far easier to include non-textual material of all types within documents. One will be able to maintain 'live links' to the system holding the original information and modify source data to update documents directly. It will be possible to distribute compound documents through electronic mail systems so that the recipient can, if authorized, edit the document on receipt. And it will be possible to retrieve documents from archival storage, incorporate and edit them into new documents and to continue the life cycle.

Glaxo's aims over the next five years are to:

- eliminate data re-entry and cut-and-paste during document production;
- improve presentation quality of documents;
- improve the company's approach to the production and management of submissions;
- improve filing, retrieval and distribution of documents;
- optimize space utilization for document storage;
- provide for seamless international exchange of documents.

This will include simple techniques for taking structures and reaction sequences into documents, a facility much needed by our research workers. We are sure that the investment in this area will produce benefits commensurate with the costs through time savings in document production and in speeding up the communication of results. And we are conscious that graphical and data systems on their own, while of tremendous benefit, are much enhanced with simple and integrated document creation techniques.

CONCLUSION

This book considers the exciting developments in chemical structure handling on which the next generations of corporate chemical information systems will be based. To bring these developments into routine operation in a commercial company requires a clear and objective view of their potential benefits and a degree of imaginative flair. As the costs of introducing IT increase, the company will examine new investment proposals more critically. Many companies are therefore concentrating at the moment on the integration of existing information systems and on improving their approach to document management, particularly when document production systems, as in the pharmaceutical industry, can speed up time to market. Justifying new chemical information systems in the future will be much, much harder. The CSA, as well as encouraging work in its traditional fields of chemical structure handling, has a role to play in fostering the development of document management systems and standards for those structures. I am sure that the CSA will continue to respond to new challenges with competence and vigour.

REFERENCES

[1] *Financial Times*, 27th January 1989. The winning mix in drug research: the spectacular success of the world's biggest pharmaceuticals group.

[2] Pharmaceutical research and development, Barclays de Zoete Wedd, May 1989.

[3] *The British Pharmaceutical Industry*, Jordan & Sons Ltd, 1986, p21.

[4] Technology moves forward in the chemical industry, *Chemical Week*, 1989, March 29th, 30–43 (144/13).

POSTCRIPT

I am going to finish this chapter on a solemn note – this is my only relevant opportunity to pay tribute to an enormously important member of the CSA who died at the very end of 1988. The CSA have already 'immortalized' Ernie Hyde with the initiation of the CSA award for contributions to chemical information work in 1978, when Ernie retired from ICI. I hope that as this award is presented at the AGM each year Ernie's achievements will be mentioned so that his memory is kept fresh by his information colleagues.

Ernie had many superb qualities and I can only mention a few of these which are particularly relevant to this book. He was a man of enormous vision and great practicality. This meant that he had the ability to welcome and plan change and also to achieve it; hence his pioneering developments of CROSSBOW and compound and biological data handling developments at ICI and his provision of the stepping stones from WLN to graphics for many companies through the release of the Daring software by Fraser Williams.

As a conference organizer and chairman, particularly in summing up, he was superb; he could see through the dross of many papers, extract the key themes, bring them together and add his own particular flair and creativity to produce real messages: ones that made you want to rush back to the office to start new projects, having absorbed some of his energy and his common sense.

Lastly – and very, very importantly – Ernie encouraged the exchange of ideas. He was a tremendous agent of 'cross-fertilization' – very generous at sharing his own experience, extraordinarily able at making others share theirs (and he had an enormous range of contacts to call on). He was open, he wanted to develop younger members and newer members of the profession; and he was approachable, very easy to get on with and enormous fun. There are a number of people who owe a considerable proportion of their success to him.

Three-dimensional structure handling

Storage and searching of stereochemistry in substructure search systems

John M. Barnard, Barnard Chemical Information Ltd., 46 Uppergate Road, Stannington, Sheffield S6 6BX, UK.
Anthony P.F. Cook, Orac Ltd., 175 Woodhouse Lane, Leeds LS2 3AR, UK.
and
Bernhard Rohde, Ciba-Geigy AG, CH-4002 Basel, Switzerland.

INTRODUCTION

To an extent this paper is out of place in this book, since it is concerned precisely with the subject which the book as a whole is trying to get beyond, namely the 2-D structure diagram and its computer representation and manipulation. However, there is some value in considering the extent to which 3-D information can be included in an essentially 2-D representation, and in discussing the methods that have been devised for doing so.

In more mathematical, or graph-theoretical language, the problem is one of using the topological representation of a chemical structure to describe its topography. The conventional structure diagram (neglecting wedge bonds etc.) represents only the topology of a molecule. It shows the chemical nature of each atom, and the connections between them, with bond orders. The connection table is simply a tabular representation of this information, and Fig. 1 shows a simple example.

Because of the very regular nature of chemical systems, and the very simple fundamental units involved (104 elements, of which only about six occur with any significant frequency – at least in organic compounds – and three bond types), this is in many cases quite adequate to describe the molecule more or less totally, or at least unambiguously. There are standard bond lengths and angles, and given the topology of a molecule, it is in principle possible to calculate the relative positions of the atoms in 3-D space. Some of the other chapters in this book discuss the problems of acheving this.

However, as every undergraduate chemist soon realizes, two molecules which have exactly the same topology can have different topographies, and cannot be inverconverted without breaking and re-forming bonds. In Fig. 1, those points in the structure diagram which can occur in more than one 3-D configuration are shown ringed around.

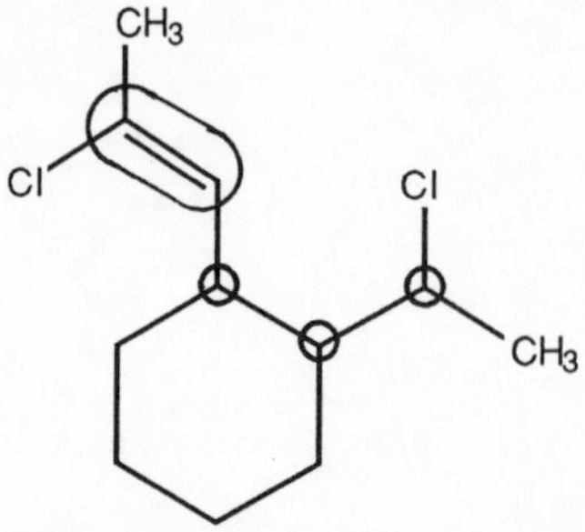

1	C	2 (s)	6 (s)	7 (s)
2	C	1 (s)	3 (s)	
3	C	2 (s)	4 (s)	
4	C	3 (s)	5 (s)	
5	C	4 (s)	6 (s)	
6	C	5 (s)	1 (s)	10 (s)
7	C	1 (s)	8 (s)	9 (s)
8	Cl	7 (s)		
9	C	7 (s)		
10	C	6 (s)	11 (d)	
11	C	10 (d)	12 (s)	13 (s)
12	Cl	11 (s)		
13	C	11 (s)		

Fig. 1 – A structure diagram and the corresponding connection table.

The different forms of stereochemistry which can occur in chemical structures have been discussed in a variety of textbooks (for example, [1–3]), and will not be reviewed here. This chapter begins with an historical overview of the methods which have been devised for indicating stereochemical information in a connection table. The use of such information in full structure and substructure searching is then discussed, along with the various ways of inputting it to the computer. In conclusion, the proposals which have recently been developed for showing stereochemistry in the Standard Molecular Data Format (SMD Format) are described.

CONNECTION TABLES AND THE MORGAN ALGORITHM

By the mid-1960s, the idea of the connection table as a machine-readable representation of the topology of a molecule was well-established. Morgan's algorithm for reproducibly assigning unique numbers to the atoms in a connection table was published in 1965 [4], and as it is of some importance in stereochemical representations, its operation will be briefly reviewed.

The algorithm has two distinct stages. The first stage attempts to classify (or partition) the atoms in the structure, effectively by examining the environment around each atom in progressively larger and larger circles. The way this is done is by assigning an initial value to each atom – the number of non-hydrogen connections which that atom has – and then at each iteration calculating a new value for each atom by adding together the values of its neighbours. The value at each atom, which is called the extended connectivity value, thus takes into account a larger and larger environment with each iteration, and the whole process is an example of the mathematical technique known as relaxation. Fig. 2 shows the extended connectivity values at various iterations for the same structure as Fig. 1.

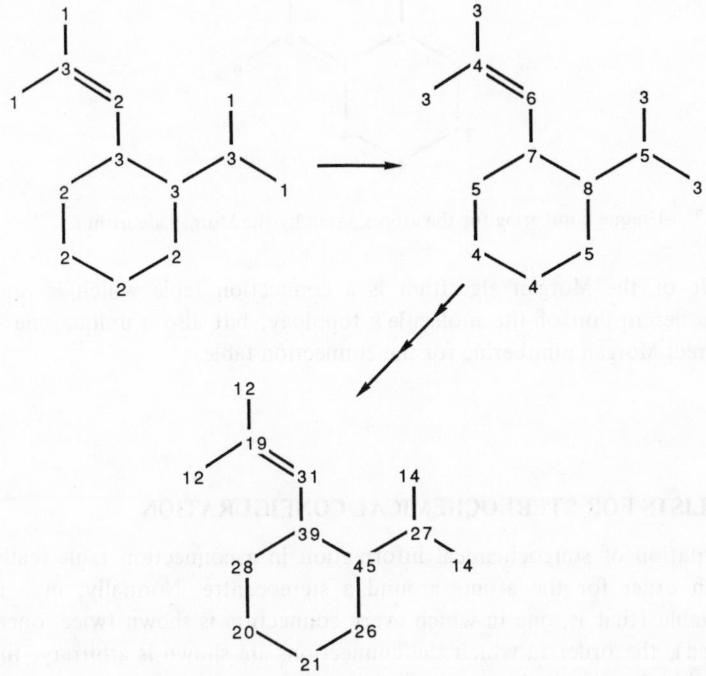

Fig. 2 – Extended connectivity values in the Morgan algorithm (explanation in text).

It is evident that a relatively small number of different atom values will exist at the first stage (just 1, 2, 3 and 4 for 1-, 2-, 3- and 4-connected atoms respectively), but the number of such atom values will normally increase at each iteration. Atoms which are symmetrically equivalent have identical environments and will therefore always have identical extended connectivity values. The first stage of the algorithm can be terminated when further iterations do not lead to an increase in the number of different values present, though in a few problem structures this stage fails to distinguish between non-symmetrical atoms.

The second stage of the algorithm involves numbering the atoms sequentially from 1 to *n*, where *n* is the number of atoms in the structure. The atom which finishes up with the highest extended connectivity value in the first stage is chosen as atom 1, its neighbours are then assigned numbers 2, 3 etc. (choosing the atom with the highest extended connectivity first), and the unnumbered neighbours of atom 2 are then numbered on the same basis. Where a choice has to be made between atoms with the same extended connectivity value, the choice is made on the basis of atom type, bond type, charge etc. Where it is still not possible to distinguish, the atoms are usually symmetrically equivalent, and it does not matter which is chosen first. Fig. 3 shows the Morgan numbering which is derived from the final set of extended connectivity values in Fig. 2.

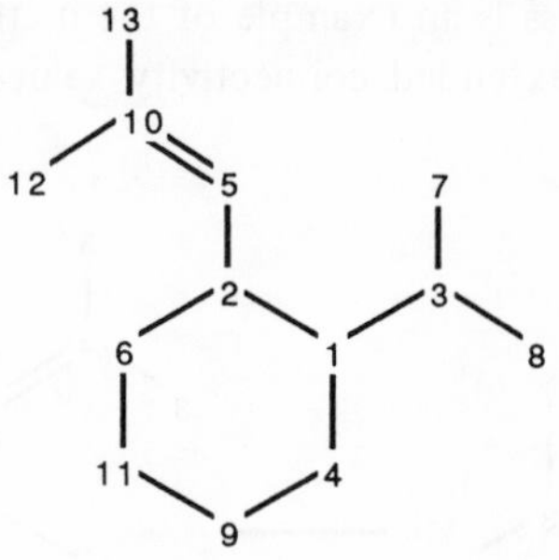

Fig. 3 – Unique numbering for the atoms, given by the Morgan algorithm.

The result of the Morgan algorithm is a connection table which is not only an unambiguous description of the molecule's topology, but also a unique one – there is only one correct Morgan numbering for the connection table.

ORDERED LISTS FOR STEREOCHEMICAL CONFIGURATION

The representation of stereochemical information in a connection table really involves specifying an order for the atoms around a stereocentre. Normally, in a redundant connection table (that is, one in which every connection is shown twice, once for each atom involved), the order in which the connections are shown is arbitrary. In a unique connection table, it might be in numerical order of the connected atoms.

If there are four connections (as around a tetrahedral carbon atom), simple permutation theory shows that there are 4! ($4 \times 3 \times 2 \times 1$) or 24 ways of arranging them. Fig. 4 shows the 24 permutations of the numbers 1, 2, 3 and 4 which are assigned to groups shown as A, B, C and D.

Because of the symmetry involved in a tetrahedral carbon atom, these 24 permutations actually fall into two classes of equivalent permutations, which are shown in the two columns of Fig. 4 (corresponding to the two possible configurations of a tetrahedral carbon). If pairs of numbers in any of these permutations are swapped round, then an even number of swaps will yield another permutation from the same column (i.e. the same configuration), whilst an odd number of swaps will yield a permutation from the other column.

The CIP rules

The well-known Cahn, Ingold, Prelog sequence rule, first proposed in the 1950s but subsequently revised [5, 6], allows a priority, based largely on atomic numbers, to be assigned to each of the groups A, B, C and D, according to the configuration of the central atom. If priorities 1, 2, 3 and 4 are then listed in the order A, B, C, D, the resulting permutation will be in the left-hand column of Fig. 4 if the central atom has the 'R' configuration, and in the right-hand column if it has the 'S' configuration.

This is not quite the way in which the R and S configurations are usually defined, but is equivalent.

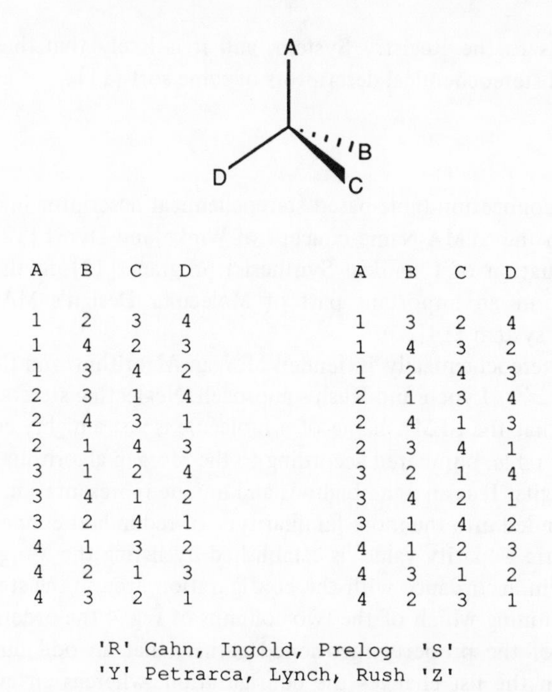

A	B	C	D	A	B	C	D
1	2	3	4	1	3	2	4
1	4	2	3	1	4	3	2
1	3	4	2	1	2	4	3
2	3	1	4	2	1	3	4
2	4	3	1	2	4	1	3
2	1	4	3	2	3	4	1
3	1	2	4	3	2	1	4
3	4	1	2	3	4	2	1
3	2	4	1	3	1	4	2
4	1	3	2	4	1	2	3
4	2	1	3	4	3	1	2
4	3	2	1	4	2	3	1

'R' Cahn, Ingold, Prelog 'S'
'Y' Petrarca, Lynch, Rush 'Z'

Fig. 4 – The 24 possible permutations of four neighbours of a tetrahedral carbon atom.

Petrarca, Lynch and Rush

Probably the earliest example of the inclusion of stereochemical information in a connection table on this sort of basis is the work of Tony Petrarca, Mike Lynch and Jim Rush at Chemical Abstracts Service [7]. They acknowledged a similarity between their work and some earlier proposals by McDonnell and Pasternack at the US Patent Office [8], who had suggested a stereochemical extension to the Hayward line notation to handle coordination compounds.

Petrarca, Lynch and Rush assigned priorities to groups A, B, C and D of Fig. 4 on the basis of the atom numberings given by the Morgan algorithm. Asymmetric carbon atoms had an extra four-integer stereodescriptor at the end of their connection table rows, in which the Morgan numbers for the connected atoms were cited in the order A, B, C, D. This enabled the configuration of the central atom to be decided, though because the way of assigning priorities to the four groups is different from the CIP rules, Petrarca, Lynch and Rush identified the two configurations as 'Y' and 'Z', and the appropriate designator was attached to the central atom. Unfortunately, there is no simple correspondence between 'R' and 'S' and 'Y' and 'Z'.

They applied a similar approach to geometrical isomerism around a double bond, though by requiring the double bonded atom with the higher Morgan number to be cited first, the number of possible permutations was reduced to 8. In subsequent work, Petrarca and Rush [9] extended the principle to square planar and octahedral geometries.

For various reasons, these ideas have never been incorporated into the design of the Chemical Abstracts Registry System, and up to the present that system uses a standardized set of descriptor terms applied to the molecule as a whole [10]. CAS is planning enhancements to the Registry System, and it is likely that these will include connection-table-based stereochemical descriptors of some sort [11].

THE SEMA NAME

Implementation of a connection-table-based stereochemical descriptor in an operational system first came with the SEMA-Name concept of Wipke and Dyott [12], in the SECS (Simulation and Evaluation of Chemical Synthesis) program [13] in the early 1970s. SEMA names also form an important part of Molecular Design's MACCS structure registration and search system [14].

SEMA stands for Stereochemically Extended Morgan Algorithm, and the idea is really a development of Petrarca, Lynch and Rush's approach. Neglecting stereochemistry for a moment, we can say that the SEMA name of a molecule is just a highly compacted non-redundant connection table, numbered according to the Morgan algorithm, and expressed as a single string of digits. It is an unambiguous and unique representation of a structure, and can easily be expanded into the more familiar type of redundant connection table.

For each stereocentre a 'parity value' is established by listing the Morgan numbers of the connected atoms, in accordance with the configuration around the stereocentre, and then effectively determining which of the two columns of Fig. 4 the ordering in question is found in. Because of the property mentioned earlier, that an odd number of interchanges of numbers in the list changes the configuration, whereas an even number of interchanges leaves the same configuration, the parity value can be established by seeing how many interchanges are needed to get the Morgan numbers into ascending order. If it is odd, the parity value is 1; if it is even the parity value is 2. 0 is used for atoms which are not stereocentres, and 3 for atoms of unknown configuration. A special 'atom configuration' list, consisting of the parity values for all the atoms in the molecule, can then be added to the end of the non-stereochemical part of the SEMA name string.

A similar approach is used for the configuration around double bonds, with a 'double-bond' configuration list being included in the SEMA name. In their original paper, Wipke and Dyott also showed that the SEMA-name principle could be extended to distinguish between different conformations in a molecule.

An additional point to note in connection with SEMA names is related to the operation of the Morgan algorithm itself. It was stated earlier that the first stage of the Morgan algorithm fails to give different extended connectivity values to atoms which are topologically equivalent, though it may be, in certain cases, that two atoms can be distinguished only on the basis of the stereochemical configuration around them. The second stage of the Morgan algorithm in the SEMA name – the stage at which the Morgan numbering for the atoms is chosen – therefore takes account of the stereochemical parity values when choosing between otherwise equivalent alternatives.

STRUCTURE REGISTRATION AND SUBSTRUCTURE SEARCH

Stereochemical information can be utilized within both full structure and substructure search algorithms.

The SEMA name provides a very excellent way of dealing with the former, because it is a very compact way of representing the full structure uniquely. Full structure search is thus simply a question of matching two SEMA names. An additional advantage of the

SEMA name is that, because the stereochemical information occurs at the end of the string, a full structure match ignoring stereochemistry can be carried out by limiting the match to the first part of the SEMA name string.

Substructure search is more complicated, because in a substructure, part of the molecule is missing by definition, and the assignment of absolute stereochemical descriptors is therefore impossible. One problem lies in finding a way of inputting the stereochemical information for a query (especially where a stereocentre may also be a free site).

The principles of substructure search have recently been reviewed by Barnard [15] and by Willett [16]. At least for atom-centred stereocentres, a simple extension can be made to the atom-by-atom backtrack search, which normally forms the final stage of a substructure search system: for a pair of atoms to match, the configuration around the central atom must be the same in both the file structure and the query structure, and this can be checked by using an appropriate ordering for the neighbouring atoms. If there are many stereocentres, this additional test may slow down the atom-by-atom searching considerably, on account of the large amount of backtracking required, though an intermediate search stage or preliminary test may be able to eliminate candidates that will not match.

Normally, stereochemical information is utilized only at the atom-by-atom search stage; the fragments used for preliminary screening do not contain any stereochemical information. In principle they could do so, though it would probably not lead to a very great improvement in retrieval performance, at least when compared with the increase in the number of fragments that would be needed.

A stereochemical substructure search system has recently been developed by Rohde [17], and this includes the use of screens with stereochemical information, which can be chosen in the light of the query types being submitted to the system.

Some of the more recently developed substructure search systems discussed by Barnard [15], which use an approach based on an hierarchical classification of the atoms (e.g. CIS SANSS, HTSS, S4, and to a lesser extent DARC), are also amenable in principle to the inclusion of stereochemical information, as an additional basis for classification.

INPUT REPRESENTATIONS OF STEREOCHEMISTRY

A major problem in the computer handling of stereochemistry, at least in graphics-based systems, is showing the configuration around particular stereocentres clearly and unambiguously.

Wedge/hash bonds

Most computer systems using graphical structure input employ some sort of variation on the wedged/hashed bond convention frequently used in structure diagrams. The problem with this type of representation is that the convention is very far from being clearly defined, and it is possible to show things with it which may be ambiguous, meaningless or even geometrically impossible.

There is also the question of whether or not the complete failure to specify stereochemistry around a particular centre is the same as the specification of a particular bond as being in 'either' configuration, or an indication that the compound being shown is a racemic mixture rather than a pure enantiomer. Whilst it is possible to show an assymetric carbon atom without stereochemical indicators, it is very difficult to show a double bond without apparently implying one or other configuration around it.

Operational computer systems, such as MACCS, define their own rules about how wedged bonds are to be used and interpreted, but unfortunately the MACCS rules do not necessarily coincide with the rules used by other systems. It does seem to be generally agreed that if two wedged bonds or two hashed bonds are shown at a single atom, they should be on opposite sides of it, whereas if there is one wedged and one hashed bond, they should be adjacent.

CIP descriptors

It may also be possible for the user to assign absolute stereochemical designators, such as the CIP descriptors, to particular atoms or bonds in a structure. This can, of course, only be done for full structures, and not for substructures, since the CIP rules require the entire molecule to be taken into account when assigning priorities. There are a number of computer programs which are able, with a greater or lesser degree of success, to apply the CIP rules automatically [18], and to calculate the correct CIP descriptor from a particular pattern of wedge/hash bonds.

Unfortunately, the CIP rules were not really designed for computer implementation, and are not particularly well suited to it; there are also a few problem structures where the rules, at least in their present published form, are unable to distinguish between certain substituents. Custer [19, 20] has shown some examples, and has also discussed a problem (the so-called 'reconstruction' problem) where two different structures can have the same set of CIP descriptors.

STEREOCHEMICAL REPRESENTATION IN THE STANDARD MOLECULAR DATA (SMD) FORMAT

During the past two years there have been moves to try to establish an internationally agreed standard connection table format, which can be used for the exchange of data [21–23].

One such format is being based on the existing 'Standard Molecular Data Format' (SMD Format) which was developed in a co-operative venture by the European chemical and pharmaceutical companies involved in the CASP group [24, 25]. There are a number of areas in which the original format requires extension or modification, and to this end a series of technical working groups has been set up to look at particular aspects of the format. Amongst these is a Stereochemistry Group, of which the present authors are all members, and the remainder of this paper discusses some of the interim proposals of the group.

An SMD Format connection table is organized as a number of separate blocks, each containing different information. Amongst the blocks available are ones for atom types, bonds, hydrogen counts, charges, stereochemistry etc. It is intended that SMD Format should also be able to represent generic structures, query structures, reactions, polymers and macromolecules, with facilities for shortcuts and sub-connection tables, which may be especially useful in the representation of generic structures and of polymers and macromolecules.

Stereochemistry in SMD Format

It has been recognized that several different means of representing stereochemistry will be required; a particular structure representation in SMD Format will be able to use any or all of these. The stereochemistry blocks which have been identified include CIP descriptor blocks, a bond picture block, and special 'rotational list' block.

Fig. 5 shows the proposed SMD format connection table for a simple substituted alkane. Separate blocks are used for the atom types, bond orders and hydrogen counts. It should be noted that two of the hydrogens are given explicit entries in the connection table (in order to allow their wedge bonds to be shown), whilst those which are attached to the hydroxyl oxygẽns are not.

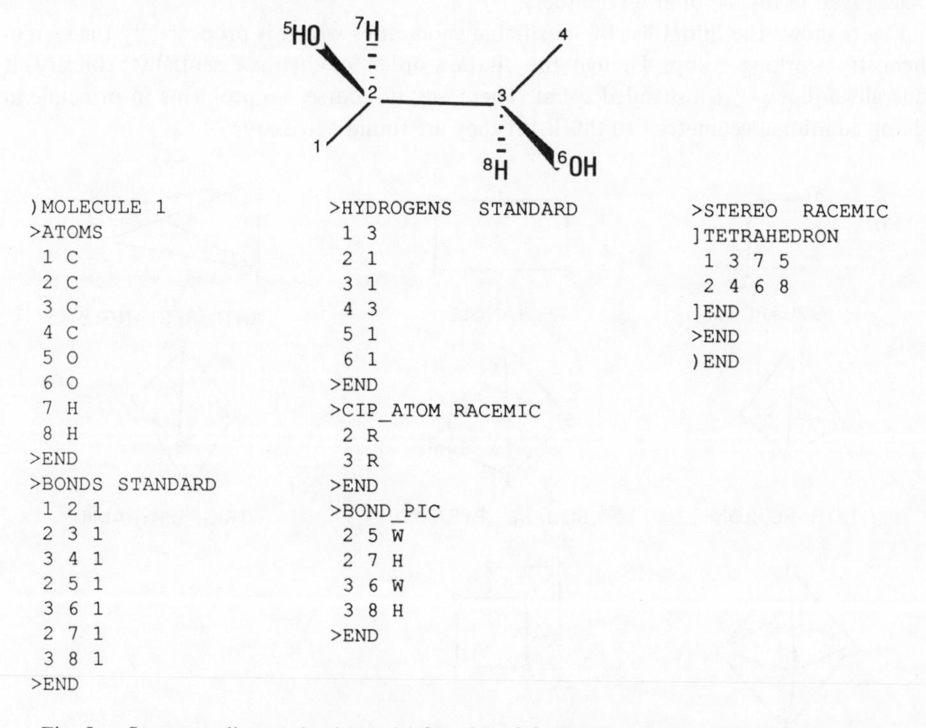

Fig. 5 – Structure diagram for (racemic) 2*R**,3*R**-2,3 dihydroxybutane, with the proposed SMD Format connection table.

In the CIP blocks, the appropriate CIP descriptors can be applied to the relevant atoms or bonds, and in the bond picture block, particular bonds can be identified as wedged, hashed or undefined. For the time being, no rigid rules will be applied to the use of wedged or hashed bonds, though it is recognized that this may cause some problems.

Rotational list description

The rotational list description is one which was proposed more-or-less independently by two of the present authors (APFC and BR) in working papers for the stereochemistry working group. It has the advantages of being fairly compact, readily processed automatically, and adjustable to almost any kind of stereo element.

The idea of the description is that, for each possible geometry, there is a specified standard order for the citation of the attachments around the stereocentre. The atom numbers for the relevant atoms (using whatever arbitrary atom numbering happens to have been used for the connection table) are simply cited in this order. There are various header lines which indicate the geometry in question, and whether the configuration being described is absolute, relative or racemic. The STEREO block in Fig. 5 illustrates the use of the method for simple asymmetric carbon atoms. In this case the relevant hydrogens have been given explicit entries in the connection table, and can therefore be referred to by number (7 and 8 in this case). It is not essential for them to be shown explicitly, even if they are involved in a stereo element. An H symbol can be used as a 'place-saver' in the list of atom numbers.

Fig. 6 shows the initial list of specifiable geometries which is proposed by the stereochemistry working group. Though the citation order for each is essentially arbitrary, it generally follows a right-handed spiral. There are, of course, no problems in principle in adding additional geometries to this list if they are found necessary.

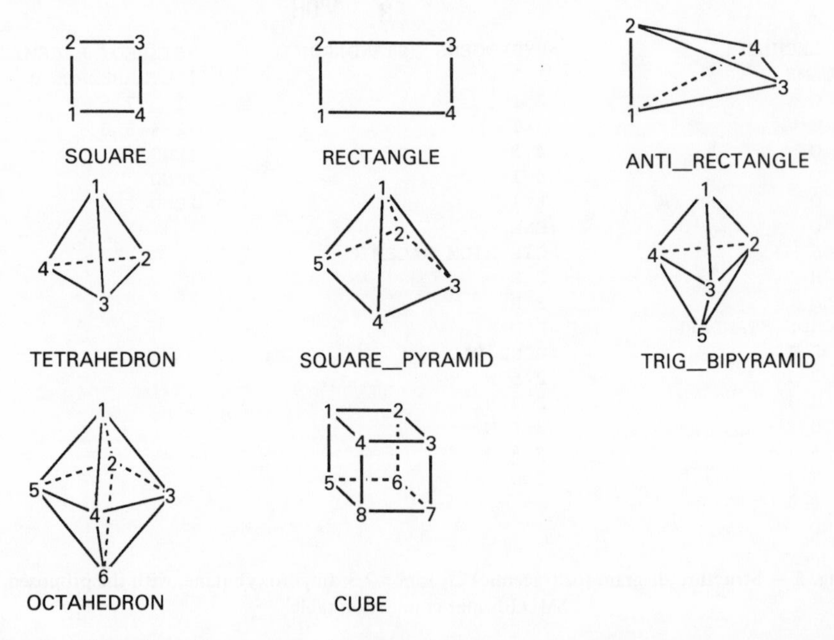

Fig. 6 – The allowable geometry keywords in SMD Format STEREO blocks, with the specified citation order for the nodes.

Fig. 7 illustrates the use of the RECTANGLE geometry to show the configuration around a double bond. In this case the colon symbol is used as a place-saver for an unshared pair of electrons. In query substructures a '*' can be used as a place-saver for a free site.

Though the intention of the working group is that this approach should be used primarily to represent stereochemical configurations, it may also be possible to use it to show particular conformations. In particular, it may be useful to indicate the arrangement of substituents in sterically hindered biphenyls, which conform essentially to the ANTI-RECTANGLE geometry. Fig. 8 shows an example.

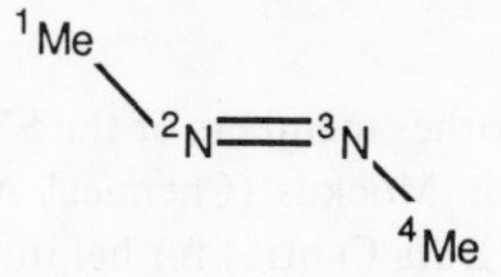

```
Trans 1,2-Dimethylhydrazine
>STEREO ABSOLUTE
]RECTANGLE
 : 1 : 4
]END
>END
```

Fig. 7 – Specification of *cis–trans* isomerism in an SMD Format STEREO Block.

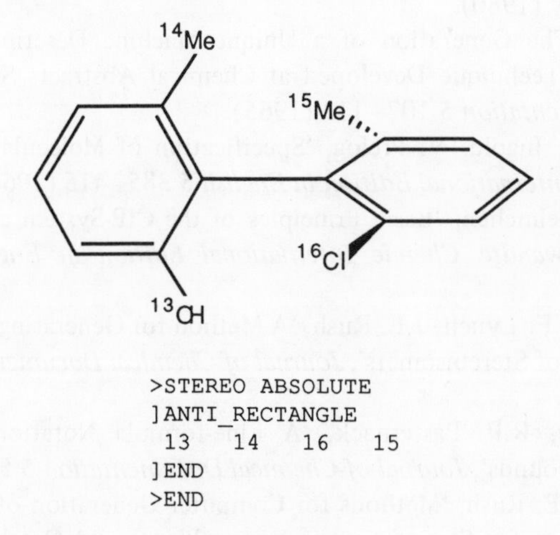

Fig. 8 – Specification of the conformation of a hindered biphenyl in an SMD Format STEREO Block.

The approach does not seem to be amenable to showing the configuration of helices or molecules that form topological knots, but the stereochemistry working group is prepared to leave these problems unresolved for the time being.

Though this type of stereochemical representation has not been described in these terms in the literature before now, it obviously has a substantial basis in earlier work, and such an approach has undoubtedly been used in many existing computer systems, at least as far as tetrahedral geometries are concerned. In fact, although the present authors were unaware of the paper when this representation was being developed, it shares many ideas with the 1965 proposals of McDonnell and Pasternack [8] for a stereochemical extension to the Hayward Line Notation. That paper included a table of specifiable geometries for coordination compounds virtually identical to that in Fig. 6, the proposal being to cite the notation symbols for the ligands around the central atom in the order specified for each geometry.

ACKNOWLEDGEMENTS

The authors would like to thank the other members of the SMD Stereochemistry Working Group, Hartmut Braun (Roche), Joe Mockus (Chemical Abstracts Service) and David Watson (Cambridge Crystallographic Data Centre) for helpful discussions. Thanks are also due to Jim Dill (Molecular Design Ltd) and Peter Löw (CHEMODATA GmbH) who kindly provided information about stereochemistry processing in MACCS, and in the R/S-Script program respectively.

REFERENCES AND NOTES

[1] E.L. Eliel, *Stereochemistry of Carbon Compounds*, McGraw Hill (1962).

[2] O.B. Ramsay, *Stereochemistry*, London: Heyden (1981).

[3] F.A. Cotton, G. Wilkinson, *Advanced Inorganic Chemistry: a Comprehensive Text*, 4th edition, Wiley (1980).

[4] H.L. Morgan, 'The Generation of a Unique Machine Description for Chemical Structures – A Technique Developed at Chemical Abstracts Service', *Journal of Chemical Documentation* **5** 107–113 (1965).

[5] R.S. Cahn, C.K. Ingold, V. Prelog, 'Specification of Molecular Chirality', *Angewandte Chemie International Edition in English* **5** 385–415 (1966).

[6] V. Prelog, G.. Helmchen, 'Basic Principles of the CIP-System and Proposals for a Revision', *Angewandte Chemie International Edition in English* **21** 567–583 (1982).

[7] A.E. Petrarca, M.F. Lynch, J.E. Rush, 'A Method for Generating Unique Computer Representations of Stereoisomers', *Journal of Chemical Documentation* **7** 154–164 (1967).

[8] P.M. McDonnell, R.F. Pasternack, 'A Line-formula Notation System for Coordination Compounds', *Journal of Chemical Documentation* **5** 56–59 (1965).

[9] A.E. Petrarca, J.E. Rush, 'Methods for Computer Generation of Unique Configurational Descriptors for Stereoisomeric Square-Planar and Octahedral Complexes', *Journal of Chemical Documentation* **9** 32–37 (1969).

[10] J.E. Blackwood, P.M. Elliott, R.E. Stobaugh, C.E. Watson, 'The Chemical Abstracts Service Chemical Registry System. III. Stereochemistry'. *Journal of Chemical Information and Computer Sciences* **17** 3–8 (1977).

[11] G.G. Vander Stouw, 'Poster Session: Potential Enhancements to the CAS Chemical Registry System'. In *Chemical Structures: The International Language of Chemistry (Proceedings of an International Conference at the Leeuwenhorst Congress Center, Noordwijkerhout, The Netherlands, 31 May–4 June 1987)*, ed. W.A. Warr, pp. 221–216. Heidelberg: Springer (1988).

[12] W.T. Wipke, T.M. Dyott, 'Stereochemically Unique Naming Algorithm', *Journal of the American Chemical Society* **96** 4834–4842 (1974).

[13] W.T. Wipke, 'Computer-Assisted Three-Dimensional Synthetic Analysis'. In *Computer Representation and Manipulation of Chemical Information*, eds. W.T. Wipke, S.R. Heller, R.J. Feldmann, E.. Hyde. pp. 147–174. New York: Wiley (1974).

[14] S. Anderson, 'Graphical Representation of Molecules and Substructure Search Queries in MACCS', *Journal of Molecular Graphics* **2** 83–90 (1984).

[15] J.M. Barnard, 'Problems of Substructure Search and their Solution'. In *Chemical Structures: The International Language of Chemistry (Proceedings of an International Conference at the Leeuwenhorst Congress Center, Noordwijkerhout, The Netherlands, 31 May–4 June 1987)*, Ed. W.A. Warr, pp. 113–126. Heidelberg: Springer (1988).

[16] P. Willett, 'A Review of Chemical Structure Retrieval Systems', *Journal of Chemometrics* **1** 139–155 (1987).

[17] B. Rohde, *GM-Search. A System for Stereochemical Substructure Search*, Ph.D. Dissertation, University of Zürich (1988).

[18] Examples include the CHIRON program from the University of Montreal, Canada, and the R/S-Script program, prcduced by CHEMODATA GmbH, D-8038 Gröbenzell, FRG.

[19] R.H. Custer, *An Investigation of the CIP System, a Mobile Molecular Model, and Computer Implementations*, Ph.D. Dissertation, Swiss Federal Institute of Technology, Zürich (1985).

[20] R.H. Custer, 'Mathematical Statements about the Revised CIP-System', *MATCH* **21** 3–31 (1986).

[21] J.M. Barnard, 'Towards a Standard Interchange Format for Chemical Structure Data'. In *Online Information 88. Proceedings of the 12th International Online Information Meeting, London 6–8 December 1988*. pp. 605–609. Oxford: Learned Information (1988).

[22] J.S. Garavelli, 'The Effort to Define a Standard Molecular Description File Format'. In *Chemical Structure Information Systems. Interfaces, Communication and Standards*, ed. W.A. Warr, *ACS Symposium Series* 400, pp. 118–124. Washing: American Chemical Society (1989).

[23] J.M. Barnard, 'Standard Representations for Chemical Information'. In *Proceedings of the Montreux 1989 International Chemical Information Conference, Montreux, Switzerland, 26–28 September 1989*, Heidelberg: Springer-Verlag (1989).

[24] Further information on SMD Format is available from the Technical Secretary of the SMD Format Group, which currently operates as a sub-group of the Chemical Structure Association: Dr J.M. Barnard, BCI Ltd, 46 Uppergate Road, Stannington, Sheffield S6 6BX, UK.

[25] H. Bebak, C. Buse, W.T. Donner, P. Hoever, H. Jacob, H. Klaus, J. Pesch, J. Römelt, P. Schilling, B. Woost, C. Zirz, 'The Standard Molecular Data Format (SMD Format) as an Integration Tool in Computer Chemistry. *Journal of Chemical Information and Computer Sciences* **29** 1–5 (1989). [Also published in *Chemical Structure Information Systems. Interfaces, Communication and Standards,* ed. W.A. Warr, *ACS Symposium Series* 400, pp. 105–117. Washington: American Chemical Society (1989)].

MACCS-3D: a new database system for three-dimensional molecular models

Thomas E. Moock, Brad Christie, Douglas Henry, Molecular Design Limited, 2132 Farallon Drive, San Leandro, California 94577.

The community of users of chemical information management software have long been used to representing chemical structures as topological, or 2-D, entities. This has changed substantially in recent years, as chemists from a variety of different domains become exposed to and use computational chemistry software. Chemists from all disciplines are using and manipulating chemical structures more and more as full three-dimensional entities, and are paying increasing attention to issues of conformation and flexibility. Fig. 1 outlines a number of chemical disciplines that are using computational chemistry programs and the results of this software.

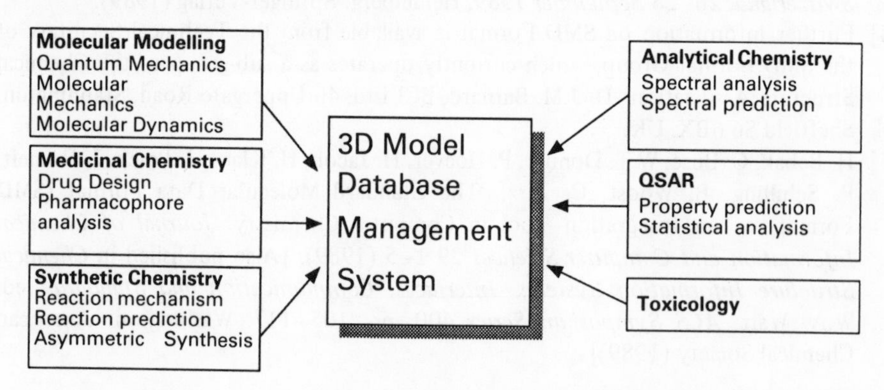

Fig. 1 – Some of the domains of chemical research now using 3-D models and modelling software.

Computational chemists, of course, are among the most prolific producers and users of 3-D models. Their numbers are growing, and with the increasing sophistication of their computational software and the increasing power of their hardware, the number of models they create is expanding rapidly. The ability to organize and fully exploit the profusion of molecular models they produce is currently an unanswered need.

Medicinal chemists were among the first to exploit topological structure databases in drug design. Their use of these databases was in part due to a desire to 'mine the corporate garbage can': that is, to test with a new assay some of the many compounds synthesized for other purposes. All biologically active compounds have side effects as well as intended ones; those found unsatisfactory in one sense may in fact make excellent lead compounds when tested for another use. The concept of the pharmacophore, and especially its geometric components, is now well-established, and in many cases the geometries of receptor sites are known with precision.

Synthetic chemists as well are using computational chemistry tools in their work, and in the most demanding endeavours have replaced physical Dreiding and CPK models. Computer programs for the simulation of chemical reactions, such as in many retro-synthesis programs, must use 3-D models and examine geometric interactions in order to predict the outcome of stereochemical transformations.

Analytical chemists are paying increasing attention to 3-D structure. Obtaining 3-D structures via crystallographic techniques is now an almost automatic process for small molecules. The use of two-dimensional nuclear magnetic resonance (NMR) techniques to study conformation has become widespread.

Chemists involved in quantitative structure–activity relationship (QSAR) research, using a variety of statistical techniques, are using 3-D models and descriptors to a greater degree than ever before. Further, there is a recent trend towards the merging of statistical modelling and molecular modelling techniques.

Chemists from a variety of backgrounds are exploiting an expanding list of computational chemistry tools, generating a tremendous number of models, and a huge volume of model-related data. At the time of this writing, this information is usually stored in an unorganized architecture, usually as files in a directory. In such a form, it is impossible to execute even simple structural searches on the models, let alone a full, 3-D geometric search. At MDL we determined that there existed a substantial need for a database management system which would store, search and manipulate 3-D models and data. This system is called MACCS-3D [1], an enhanced version of Molecular Design Limited's Molecular Access System (MACCS) [2].

The design goals of MACCS-3D were

– to preserve the historical record of modelling studies, by storing the resulting models and model-related data;
– to provide a flexible set of efficient goemetric search tools, fully integrated with MACCS's topological searching system;
– to provide an open architecture for communication with other modelling tools, such as model builders and graphics display software and hardware;
– to take the next essential step in future QSAR systems, by providing a database of 3-D models for study;
– to create a graphical, interactive program accessible and usable by a large body of chemists who may be unfamiliar with computational chemistry.

Several software systems are available, or have been presented or published, at the time of this writing which accomplish some of the capabilities mentioned above. The most widely known and described is the Cambridge Structure Database (CSD), distributed by the Cambridge Crystallographic Data Centre (CCDC) [3]. Although primarily considered a vendor of crystallographic data, CCDC distributes with the database a set of software tools supporting a number of geometric searching features, and a package of statistics programs for reporting results. The system also includes software for the display of 3-D structures in a number of different formats. More recently CCDC has announced the development of a new graphical user interface for searching 2-D structures in the diagram file of CSD. As of this writing the software developed and distributed by CCDC is intended for use with CSD exclusively, and the accompanying software contains no facilities for registration of personal or proprietary models.

Martin and coworkers at Abbott Laboratories have recently developed an in-house system for registration, retrieval and searching of 3-D models and data [4]. Created as an enhancement of Daylight Systems, structural database software, the Menthor [4a] module manages the storage and retrieval of models, and the Aladdin [4b] module is responsible for searching the database with queries containing geometric constraints.

Sheridan and coworkers [5] at Lederle Laboratories have recently described another in-house system which searches databases of 3-D models with geometric constraints. The program can process queries with distances, angles, dihedral angles and volumes of exclusion.

A substantial body of research on geometric searching has recently been published by Willett [6]. Studies on geometric key searching, search algorithms for interatomic distances, and 3-D maximum common subgraph determination have been reported.

MACCS-3D DATABASE STRUCTURE

Fig. 2 outlines the structure of the MACCS-3D database. It is composed of two tightly linked registries: a 2-D *structure* database, and a 3-D *model* database. A one-to-many structure–model relationship is maintained between the two parts.

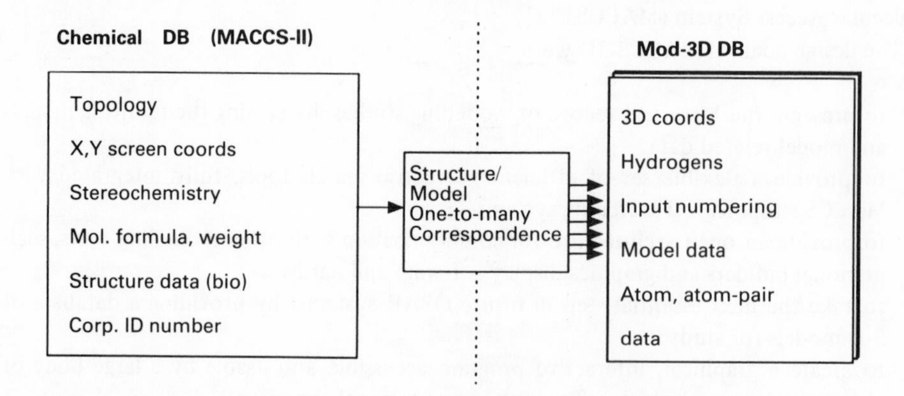

Fig. 2 – Outline of the MACCS-3D database structure. Data relating to 2-D structure resides in the chemical database, while data relating to specific conformations resides in the model database.

The structure database contains structural data such as the topology (connectivity) of the entries, low-resolution X, Y screen coordinates, stereochemistry, molecular formula and weight, and so on. This registry is the standard structure database used by MACCS-II. Any other data related to a structure (but not to a particular conformation), such as biological activity data, is stored in the MACCS-II database. Any structure may correspond to an arbitrary number of 3-D models, which are stored in the model database.

The model registry parallels the structure registry in most respects. Many commands recognized by MACCS-II to operate in the 2-D domain will also work in a consistent manner in the 3-D domain. The model database includes floating-point atom Cartesian coordinates, hydrogens (which are normally stripped during registration into the structure database), lone pairs and the original numbering of the atoms prior to registration. Important components of a 3-D database will also include data related to particular conformations, such as model source, confidence level, minimization energy, and so on. All standard MACCS-II data formats are supported in the model database, such as formatted and flexible text, numeric ranges, and so on.

In addition to data relating to a model as a whole, MACCS-3D can also store data relating to a particular atom or pair of atoms. This is important for the storage of analytical data, and data generated from computational chemistry programs: partial charges, atom coordinates within unit cells, partial bond orders, nuclear overhauser effect (NOE) data from 2-D NMR, etc. An entire distance matrix may be stored in this manner.

SEARCHING OF 3-D MODELS AND MODEL-RELATED DATA

Many of the important search capabilities are listed in Fig. 3. All data stored in a MACCS-3D database can be searched, using the same methods as in standard MACCS-II. Data pertaining to specific atoms or atom pairs is searchable either as model data, or in conjunction with a substructure search (see below).

An important component of any structural search system is the ability to establish exact or approximate identity. For topological structures, this is done in MACCS-II by generating a canonical 'name', using the Stereochemically Extended Morgan Algorithm (SEMA) [7], and using a hashing function to quickly identify duplicate structures in the database. Finding duplicate structures serves the important role of preventing the database from accumulating redundant information. Unfortunately, a hashing scheme does not work well in the model domain, since exact identity of models is rarely observed; instead, models must be judged identical only within a certain range of permissible error. For that reason MACCS-3D determines the equivalence of two models (within a permissible error) by determining the best-fit 3-D transformation matrix required to overlay one model onto the other.

The most interesting of the searching features in MACCS-3D pertain to 3-D geometric structure searches. As shown in Fig. 3, geometric structure searches can contain several components.

The most basic components are the *topological constraints*: these are the substructure fragments normally used when performing a substructure search in MACCS-II. In Fig. 3, this would be the 1,3-dihydroxycyclohexane structure. The fragments need not be connected; in fact geometric queries often contain only disconnected atoms.

- Model or per-unit data searches
- 3D structure searches
- 2D connectivity
- Geometric constraints
- Per-atom/pair data constraints
- Exact-match model search

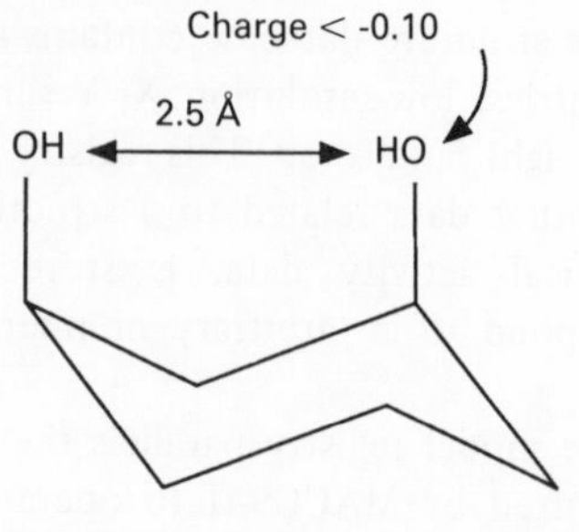

Charge < -0.10

2.5 Å

OH HO

Fig. 3 – Types of searches available in MACCS-3D for models, model-related data and atom or atom-pair data.

Upon the substructure components can also be constructed *geometric objects*. Fig. 4 shows the types of geometric objects recognized by MACCS-3D; they include points, lines, planes, centroids and normals. By default all atoms in a substructure query are predefined points, which can be used to construct other geometric objects which do not necessarily lie on the query structure itself.

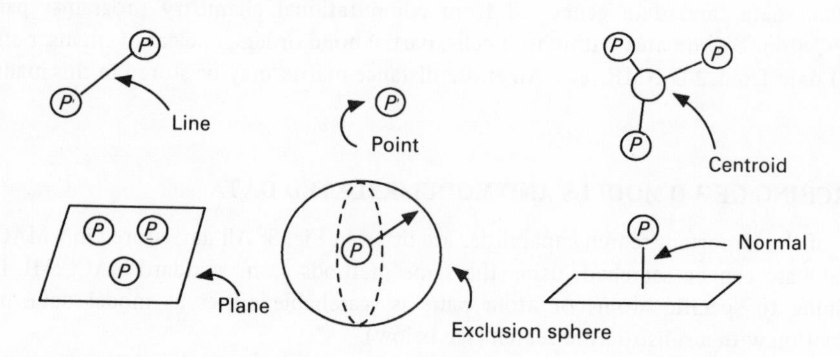

Fig. 4 – types of 3-D geometric objects available for constructing search queries.

The objects mentioned above do not in themselves determine whether a candidate entry in the database is selected by the search process. This is done by the satisfaction of a set of *geometric constraints*, which are built from geometric objects. Fig. 5 shows how tolerance values can be used to create constraints from certain objects. Two or more points can be used to define a line; if more than two points are used, MACCS-3D will construct the best-fit line through the points, and an optional *error constraint* can be used to select only models containing a roughly linear set of atoms. Similarly, if more than three atoms/points are used to construct a plane, MACCS-3D will construct the best-fit plane through the points, and an optional error constraint can be applied to select models containing planar collections of atoms.

Fig. 5 also includes an example of how several objects can be used to construct a more complex set of geometric constraints. The goal of this example is to construct an exclusion sphere above the aromatic ring, defining a query which is satisfied if no atoms lie directly above the ring. The atoms of the benzene ring are used to construct a plane and a centroid point; the centroid and plane are then used to construct a vector normal to

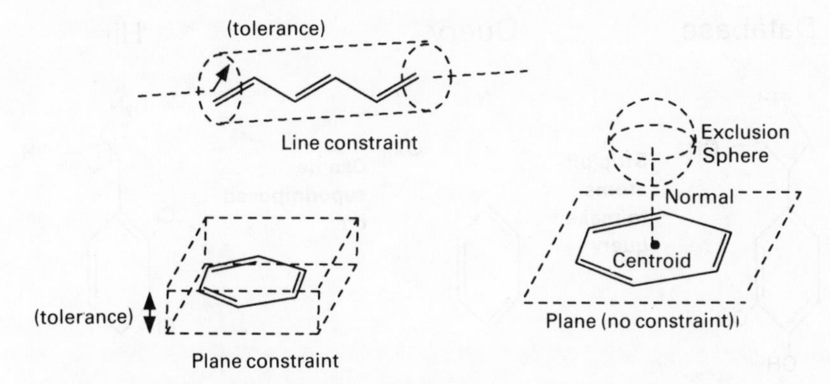

Fig. 5 – Use of lines and planes with error constraints. Also, an example of a complex set of objects used to create an excluded volume constraint over an aromatic ring.

the plane of the ring. A new point is constructed a specified distance along the vector, which is then used to create an excluded volume constraint.

The most important geometric constraints, however, are shown in Fig. 6. These include distances, angles, and dihedral angles. Interatomic distance constraints can be defined by selecting two atoms (or points) in the query. The chemist can also construct distance constraints between other objects, such as lines and planes. Angle constraints are defined by selecting three atoms/points, which need not be directly connected; angles between lines and planes can also be specified. Dihedral, or torsional angles, can be constructed from four points, which also need not be directly connected.

Atom or atom-pair data can also be used to constrain a geometric search, as indicated previously in Fig. 3. Thus a requirement of a partial charge at a specific atom can be used in conjunction with a substructure and/or geometric query.

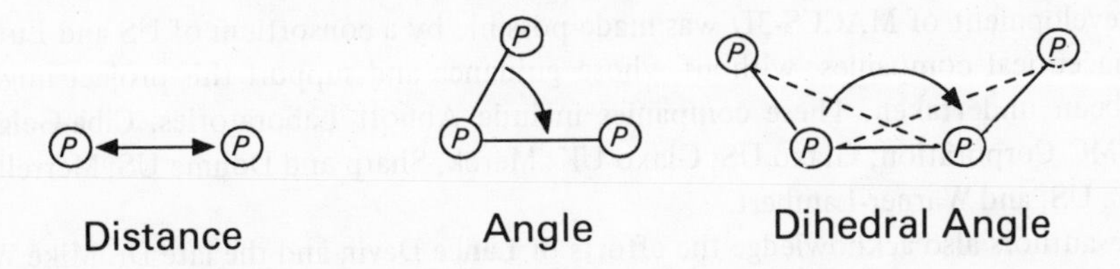

Fig. 6 – Geometric constraints available in MACCS-3D for constructing search queries.

In addition to the objects and constraints mentioned above, MACCS-3D can also use as a query a 3-D model or fragment of a 3-D model, preserving the 3-D relationship of its component atoms. This is shown in Fig. 7. In the queries described previously, atoms and bonds can be created freehand in the draw menu, without regard to whether the screen coordinates are accurate. A benzene ring drawn in a distorted fashion will still hit all benzene ring-containing models in the database. However, if the atoms are *fixed* (within a tolerance set by the chemist), their spatial arrangement must be precisely duplicated in the database entry for the candidate model to be selected. This method is most meaningful when an actual 3-D model, rather than a hand-drawn structure, is used as the basis of the query.

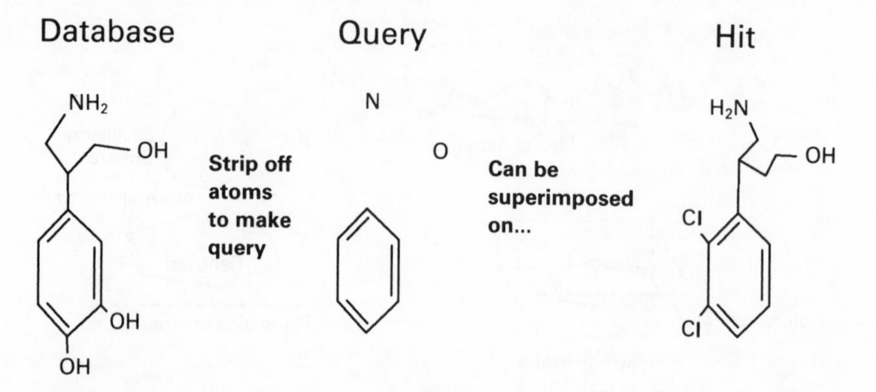

Fig. 7 – 'Fixing' the atoms in a query, usually derived from a 3-D model, will require selected models to precisely duplicate the spatial relationship of the atoms in the query.

VISUALIZATION OF SEARCH RESULTS

After completion of a search, the chemist can browse the selected models in the searching menu. This process is facilitated by graphically overlaying the query, including all geometric constraints, onto the selected model in a highlighted colour. The range of acceptable values normally displayed on the query (e.g. '4.5–5.0 Å') is replaced by the values measured on the selected model (e.g. '4.95 Å'). Tools for making arbitrary measurements of distances and angles are also available to the chemist.

ACKNOWLEDGEMENTS

The development of MACCS-3D was made possible by a consortium of US and European pharmaceutical companies, without whose guidance and support this project might not have been undertaken. These companies include Abbott Laboratories; Ciba-Geigy US; The FMC Corporation; Glaxo US; Glaxo UK; Merck, Sharp and Dohme US; Merrell-Dow, Sandoz US, and Warner-Lambert.

The authors also acknowledge the efforts of Lance Devin and the late Dr. Mike Wagner (customer and consortium communications), Kim Wiseman-Sleeter (documentation), Radford Low (lead test engineer), and Mita Imam (database development).

The MACCS-3D system is based on ideas first developed by W. Todd Wipke.

REFERENCES

[1] Wipke, W.T., 'Three Dimensional Substructure Search', presented at the 186th National Meeting of the American Chemical Society, Washington D.C., August 1983.

[2] (a) Dill, J.D., Hounshell, W.D., Marson, S., Peacock, S., Wipke, W.T. 'Search and Retrieval Using an Automated Molecular Access System', presented at the 182nd National Meeting of the American Chemical Society, New York, August, 1981.

(b) Ahrens, E.K.F., 'Customisation for Chemical Database Applications'. In *Chemical Structures: the International Language of Chemistry*, Warr, Wendy A., Ed., Springer-Verlag, London, 1988, pp. 97–111, and references cited therein.

[3] Allen, F.H., 'The Cambridge Structural Database as a Research Tool in Chemistry'. In *Modelling of Structure and Properties of Molecules.*, Maksić, Z.B., Ed.; Ellis Horwood Limited, Chichester, 1987; pp. 51–66, and references cited therein.

[4] (a) Martin, Y.C., Danaher, E.B., May, C.S., Weininger, D., 'MENTHOR, a Database System for the Storage and Retrieval of Three-dimensional Molecular Structures and Associated Data Searchable by Substructural, Biologic, Physical or Geometric Properties', *J. Comput. Aided Mol. Design* 1988, **2**, 15–29.

(b) Martin, Y.C., Danaher, E.B., May, C.S., Weininger, D., Van Drie, J.H., 'Strategies in Drug Design Based on 3-D-Structures of Ligands'. In *QSAR: Quantitative Structure–Activity Relationships in Drug Design*, Fauchère, J.L., Ed., Alan R. Liss, Inc., New York, 1989, pp. 177–181.

[5] Sheridan, R. P., In a paper presented at the Drug Information Association Meeting, San Francisco, January 1989.

[6] Brint, A.T., Mitchell, E., Willett, P., 'Substructure Searching in Files of Three-Dimensional Chemical Structures'. In *Chemical Structures: The International Language of Chemistry*, Warr, Wendy A., Ed., Springer-Verlag, London, 1988, pp. 131–144, and references cited therein.

[7] Wipke, W.T., Dyott, T.M., 'Stereochemically Unique Naming Algorithm'. *J. Amer. Chem. Soc.*, 1974, **96**, 4825–4834.

Graph matching techniques for databases of three-dimensional chemical structures

Helen M. Grindley, Eleanor M. Mitchell, Peter Willett, Department of Information Studies, University of Sheffield.
Peter J. Artymiuk and David W. Rice, Department of Molecular Biology and Biotechnology, University of Sheffield.

1 INTRODUCTION

Computer-based information systems have long played a vital role in chemical research. As well as supporting traditional numeric or textual files containing, for example, biological test data or laboratory reports, these systems also provide sophisticated facilities for the storage and retrieval of information pertaining to chemical structures [6]. A molecule in such an information system is usually represented as a 2-D chemical structure diagram, the *lingua franca* of chemists the world over, using a data structure known as a *connection table*. This contains a list of all of the non-hydrogen atoms within a structure, together with bond information that describes the exact manner in which the individual atoms are linked together, i.e., the *topology* of the molecule. The location of the hydrogen atoms within a molecule can be deduced from the bond orders and atomic types of the non-hydrogen atoms. An important characteristic of a connection table is that it can be regarded as a *graph*, a mathematical construct that describes a set of objects, called *nodes*, and the relationships, called *edges*, that exist between pairs of the objects. Thus, in the case of a 2-D connection table, the atoms and bonds correspond to the nodes and edges of a graph [30].

The use of a graph-theoretic description for a molecule means that searching operations on databases of machine-readable chemical structures can be implemented using *isomorphism* algorithms, which compare one graph with another to determine the equivalence relationships that exist between them [26]. Common requirements are the ability to search a database of compounds for the present or absence of a query compound, the process of *structure searching* or *registration*, or for all of those molecules in the database that contain some user-defined partial structure, the process of *substructure searching*. These two searching tasks correspond to *graph isomorphism*, the comparison of one graph with another to determine whether they are identical, and to *subgraph isomorphism*, the identification of the presence of a query graph within a larger graph, respectively. A further type of graph algorithm that is used in chemical information systems is the *maximal common subgraph algorithm*, a computational procedure for identifying the largest subgraph common to a pair of graphs, this is used for the detection of the reaction sites in chemical reaction database systems and for structure elucidation studies *inter alia*. Until recently, the use of graph-theoretic algorithms in chemical information science has been restricted to the processing of 2-D molecules, i.e., the techniques that have been developed have all focussed on the topology of molecules, without consideration of their *topography*, the orientation of the constituent atoms in 3-D space.

The increasing availability of 3-D structure data has led to the development of sophisticated molecular graphics systems which can be used for the interactive analysis and display of the 3-D structures of chemical molecules. However, these systems have generally not possessed facilities which allow the coordinate data to be used as the basis for 3-D searching operations analogous to those available for the searching of databases of 2-D chemical molecules. This limitation provided the initial impetus for an ongoing programme of research at the University of Sheffield into graph-theoretic techniques for the storage and retrieval of 3-D chemical structures. The work commenced with studies of the use of graph matching techniques for the small molecules in the Cambridge Structural Database (CSD) [3], this leading to subsequent work on the macromolecular structures in the Protein Data Bank (PDB) [2, 7]. In both cases, we have needed to develop an appropriate form of structure representation.

The two main components of a connection table are a list of structural elements together with information about the relationships between pairs of these elements: in the case of a 2-D molecule, the structural elements are the atoms and bonds, these corresponding to the nodes and edges of a graph. In the case of a 3-D molecule, the connection table must describe not actual physical linkages but the relative positions of the individual structural elements; accordingly, the entries in the main body of the table contain geometric information that describes the angular and distance relationships between pairs of such elements. We have used two different types of connection tables for our work with the CSD and with the PDB:

- In the case of the small 3-D molecules in the CSD, the nodes and edges of a graph correspond to the non-hydrogen atoms and to the interatomic distances respectively. An entirely analogous representation can be used to describe the geometric arrangement of the Cα atoms in a protein (although this is obviously very much larger owing to the great number of residues in the typical PDB structure).
- In the case of our extended studies of techniques for the processing of protein *secondary structures* [4], we have made use of the fact that the two most common types of secondary structure element, or SSE, are the α-helix and β-strand and that these are both approximately linear repeating structures which can hence be described by a vector drawn along their major axes [1, 27]. The set of vectors corresponding to the SSEs in a protein can then be used to describe the structure of that protein in 3-D space, with the SSEs and the inter-SSE angles and distances corresponding to the nodes and to the edges of a graph, respectively. Mitchell *et al.* describe the sequence of operations needed to generate such a connection table from the coordinate data for a PDB structure [23].

An obvious consequence of the use of these two types of 3-D representation is that there is an edge linking each and every pair of nodes in the graph that is used to represent a molecule: the chemical graphs considered here are thus examples of *fully connected graphs* (although 'wild-card' facilities are available which permit the searching of query substructures in which edges are specified only between certain of the nodes in the query graph [24]). Accordingly, if the connection table represents a molecule with NS structural elements (either atoms or SSEs), the main body of the table contains $NS(NS-1)/2$ distinct entries and any algorithm for the processing of the entries in this table is likely to have a time complexity of at least $O(NS^2)$ (in fact, the complexity is normally very much greater as we shall see below). In the case of a 2-D molecule, conversely, the great majority of the atoms are bonded to not more than four other atoms, so that the corresponding algorithms here have a minimal time complexity of $O(NS)$. It is thus to be expected that the processing of 3-D chemical graphs is likely to be much more demanding of computational resources than is the processing of 2-D chemical graphs.

Having discussed the representation of 3-D chemical substances by means of graphs, the remainder of this paper outlines the uses we have made of subgraph isomorphism and maximal common subgraph isomorphism algorithms for the processing of databases of 3-D molecules; in addition, we also describe approaches to the calculation of intermolecular structural similarity.

2 SUBGRAPH ISOMORPHISM ALGORITHMS

2.1 The Ullman algorithm

The obvious way to check for the presence of some NQ-atom substructure in a larger NS-atom database structure, a procedure that is usually referred to as *atom-by-atom searching*, is to generate each of the possible combination of NQ atoms from the database structure and then to determine whether one of the $NQ!$ permutations of each such combination is isomorphic with the query. There can be as many as $NS!/(NS-NQ)!$ possible combinations and subgraph isomorphism is thus an example of an *NP-complete* problem, i.e., one for which no polynomial time algorithm exists (or, indeed, can exist). Accordingly, since substructure searching involves checking for the presence of a subgraph isomorphism between the query substructure and each of the molecules in a database, it is highly demanding of computational resources.

Gund [17] seems to have been the first person to suggest that atom-by-atom searching techniques for databases of 2-D molecules could also be used for the automatic identification of pharmacophoric patterns in 3-D molecules, a procedure which is referred to by Jakes *et al.* as *geometric searching* [18]. Willett and his co-workers have reported comparative studies of the efficiencies of a range of subgraph isomorphism algorithms when they are used for geometric searching [8, 9]. The most generally useful was found to be one due to Ullman [31] (which has been shown subsequently to be closely related to the *relaxation* algorithm for 2-D chemical substructure searching [15]).

The Ullman algorithm does not include an explicit stage for the generation of combinations of atoms. Instead, it operates by means of a backtracking tree search in which database atoms are tentatively assigned to query atoms and the match extended in a depth-first manner until a complete match is obtained or until a mismatch is detected; in this case, the search then backtracks to the previous assignment and an alternative match is considered. Backtracking search is a common technique for increasing the efficiency of graph matching algorithms: Ullman's contribution was to identify an heuristic, which he referred to as the *refinement* procedure, that limits the number of levels of the search tree that have to be investigated before a mismatch is identified. Specifically, the algorithm makes use of the fact that if some query atom, $Q(X)$, has another query atom, $Q(W)$, at some specific distance and if some database atom, $S(Z)$, matches $Q(W)$, then there must be some database atom, $S(Y)$, at the appropriate distance from $S(Z)$ which matches $Q(X)$; this is a necessary, but not sufficient, condition for a subgraph ismorphism to be present. The refinement procedure is called before each possible assignment of a database atom to a query atom; the matched structure is increased in size by one atom if, and only if, the condition holds for all of the possible values of W, X, Y and Z. Ullman advocated this particular refinement technique in the context of general graphs, without any specific application context. The algorithm seems to be particularly well suited to 3-D chemical structures since, as noted previously, the graphs are fully connected and there is thus a very large amount of information available to the refinement procedure, this resulting in the very rapid detection of query atom-to-database atom mismatches.

Work at Sheffield over the last three years has used the Ullman algorithm for three related geometric searching problems as follows:

- Searching for pharmacophoric patterns, i.e., patterns of atoms, in the small molecules of the CSD [9].
- Searching for patterns of Cα atoms in the PDB [8].
- Searching for patterns of secondary structures, or *motifs*, in the PDB [4, 23, 24].

In all cases, we have found that it provides an efficient and an effective means of detecting the presence of topographic patterns (and have also demonstrated that the refinement procedure can be used for the detection of topological patterns [15]). Accordingly, until evidence can be presented to the contrary, we would expect it to be the algorithm of choice for any chemical retrieval system that involves some sort of substructure searching component.

The main limitation of the Ullman algorithm is its storage requirement when large molecules need to be processed, as is the case with geometric searching for patterns of Cα atoms in proteins (where the graphs may well contain many tens or even hundreds of nodes). In this case, an algorithm due to Lesk [20] provides an effective pre-search screen that reduces the numbers of nodes that need to be considered in the backtracking tree search. This algorithm provides a partitioning technique that is analogous to the set reduction methods that have long been used for 2-D substructure searching [6]. Specifically, an equivalence table is constructed that lists those database structure atoms that have atoms at the same distance (to within any tolerance that is allowed) as do the atoms in the query structure. These sets of possible equivalences are iteratively refined until it is not possible to reduce the number of equivalences any further. In Lesk's original algorithm, the remaining equivalences acted as the input to the combinatorial generation procedure that has been described previously; in our work, these equivalences are instead used as the input to the Ullman aglorithm [8]. In addition to reducing the memory requirements of the Ullman algorithm, the partitioning stage can also be expected to decrease the CPU requirements owing to the very much smaller number of database atoms which need to be considered for matching against each of the *NQ* query atoms during the backtracking stage.

2.2 The POSSUM system

We shall illustrate the use of the Ullman algorithm by means of a novel 3-D searching system that is based on this algorithm. The system, called POSSUM (*P*rotein *O*nline *S*ubstructure *S*eaching – *U*llman *M*ethod), allows a user to specify a query motif, a 3-D pattern composed of α-helices and/or β-strands, and then to search for all occurrences of this motif in the PDB [24]. As noted in Section 1, the encoded PDB structures and query motifs are represented for search by graphs in which the nodes and edges correspond to the SSEs and to the interaxial angles and midpoint and closest approach distances, respectively, and these graphs are searched using the Ullman algorithm. The inclusion of three types of information in the edges of the graphs provides a range of searching options depending upon the requirements of a particular query. The current version of the search program allows the user to take one of four options when searching for a pattern:

- By setting the distance tolerance to a very large figure (e.g., 500Å), the user can search for the pattern using purely angular constraints, although experience suggests that this does not give a very specific search.
- By setting the angular tolerance to a very large figure (e.g., 360°), the user can search for the pattern using purely distance constraints.
- If the pattern is known to contain secondary structural elements that are separated by large distances (i.e., > 10Å), the user can specify a distance tolerance for the distance between the midpoints of the axes lines.
- If the distances between the elements are small (i.e., less than 10Å), the distance tolerance can be specified for the closest approach distance.

The distance tolerances can be expressed as absolute quantities, e.g., a certain number of angstroms, or as a proportion of the distance concerned, e.g., 10%. In practice, the closest approach distance has been found to be more effective for retrieving secondary structures which interact with each other in some sense, e.g., neighbouring pairs of β-strands in a sheet, or helices in van der Waals contact with each other. However, the midpoint distance appears more appropriate for the case where the secondary structure elements are more remote from one another in the structure. The user can also specify whether or not the order that the SSEs occur along the protein chain is important in the search for that motif. Matches to the query pattern are output by the program in a format compatible with the FRODO graphics package; this allows the immediate inspection of the highlighted secondary structures which match the motifs, on an Evans and Sutherland PS300 vector graphics terminal. The program is written in Fortran 77, and is implemented on a Digital Equipment Corporation MicroVAX-II system. A scan of the entire PDB typically requires about 250 CPU seconds (though the precise value depends upon the specific motif that is being searched and the error tolerances that are used).

POSSUM has been extensively tested by means of searches of the PDB for many different query motifs [4, 24]. The results demonstrate that the Ullman algorithm provides an extremely efficient means of identifying all occurrences of these motifs, occurrences which, in some cases, had not been recognized previously owing to the complexity of the proteins' structures. In 1977, for example, Richardson reviewed all of the 37 β-strand patterns then known [28]. Searches were carried out in the PDB for 34 of these motifs (the other three not being searched since they are very simple motifs that occur in nearly all strand-containing proteins). Given the extensive studies that have been carried out using the PDB and given the relatively small number of structures in it, *ca.* 400, it is remarkable to find that no less than 15 of these searches resulted in the identification of at least one occurrence of the query motif that had not been observed by Richardson [4]. As another example, we have recently established that there is a striking resemblance in the tertiary fold of the *Salmonella typhimurium* CheY chemotaxis protein and that of the GDP-binding domain of *E. coli* elongation factor Tu (EF Tu). These two protein structures are representatives of two major macromolecular classes: CheY is a signal transduction protein with sequence homologies to a wide range of bacterial proteins involved in regulation of chemotaxis, membrane synthesis and sporulation; whilst EF Tu is one of the family of guanosine nucleotide binding proteins which includes the *ras* oncogene proteins and signal transducing G proteins. The similarity revealed by POSSUM extends far beyond previously recognized resemblances of each protein's fold to that of a generic nucleotide binding domain, thus demonstrating the utility of the program for exploratory studies of protein structure [5].

2.3 Use of screening techniques

Substructure searching in 2-D chemical databases normally involves some type of *screen search* which is used to eliminate the great bulk of the database, so that only a few molecules have to undergo the time-consuming atom-by-atom subgraph isomorphism search. A screen is a substructural feature, the presence of which is necessary but not sufficient for a molecule to contain the query substructure. These features are typically small, atom-, bond- or ring-centred fragment substructures which are generated from a connection table, and the screen search involves checking each of the database structures for the presence of those screens which are present in the query substructure [18].

We have shown that analogous techniques can be used to provide dramatic increases in the efficiency of geometric searching. Thus, Jakes and Willett have discussed criteria for the selection of screens that consist of pairs of atoms together with an associated inter-atomic distance range [19]. These ranges were chosen so as to occur approximately equifrequently in the file that was to be searched, *ca.* 13 000 3-D structures from the CSD. Searches with published pharmacophoric patterns showed that more than 99% of the file could be eliminated prior to the detailed geometric searching algorithm that determines whether the query pattern is contained within a 3-D structure.

The situation with the PDB is rather different since there are far fewer structures in the database; moreover, each of these is extremely large and thus not obviously susceptible to the screen-based encoding methods that are appropriate for the characterization of small molecules. Thus, rather than trying to eliminate entire structures from the atom-by-atom search, the need here is for criteria that will allow the elimination of some of the individual components of each structure. An example of this is provided by our work on searching for patterns of Cα atoms where it may be possible to define the query atoms in terms of the residue type, e.g., polar or non-polar, and of the type of secondary structure in which the residue occurs. A given query atom then need be considered for matching only against those atoms from a PDB structure that are of the same type. The efficiency of such an approach is discussed by Brint *et al.* [8]. Our implementation of the first part of Lesk's algorithm for searching patterns of Cα atoms can also be considered as a form of screen since we again seek to eliminate many of the database atoms from further consideration prior to the geometric search: as described previously, the criterion used here is that a database atom must have other database atoms at the same distances as a query atom has other query atoms [8].

3 INTERMOLECULAR SIMILARITY PROCEDURES

3.1 Introduction

The systems described above are all based on the use of subgraph isomorphism techniques. The last few years have, however, seen increasing interest in the use of techniques for the determination of the *similarity* between pairs of chemical graphs; thus, rather than finding structures that *contain* a query partial structure, we are here interested in structures that are *related* in some quantitative way to the query. A review of similarity methods for 2-D chemical structures that have been developed in previous work in Sheffield is provided by Willett [32]. These studies have rapidly been taken up in a range of operational environments and the success of the work has led us to consider the development of analogous techniques for 3-D chemical structures. This work is discussed in the remainder of this section.

3.2 Maximal common subgraph isomorphism

The first type of 3-D similarity measure we have considered is for use with the small molecules in the CSD and is based on the use of *maximal common substructure* (MCS) algorithms. The MCS is a set of atoms, and the associated interatomic distances, that is common to the structures being compared and that is larger than any other common substructure; there may, of course, be several common substructures of the maximum size. The identification of common 3-D substructures is computationally demanding since it corresponds to the graph-theoretic problem of *maximal common subgraph* identification (which, like subgraph isomorphism, is known to be NP-complete). However, it is of great importance in medicinal chemistry since the MCS for a set of structurally disparate 3-D molecules of known biological activity is likely to represent, or at least to contain, the pharmacophoric pattern, i.e., the geometric pattern of atoms responsible for the observed activity [14].

There are two MCS algorithms available that can be applied to the matching of 3-D molecules. In work at Stanford, Crandell and Smith [14] have described a breadth-first search procedure that enables the identification of the 3-D substructures in common between a set of molecules. The algorithm involves taking all the common substructures of size N atoms associated with each molecule and adding an extra atom to each of them. These enlarged substructures are then compared. If a substructure is not found in all of the other molecules, it is deleted from consideration. The surviving substructures form the common substructures of size $N + 1$ that act as the input to the next iteration of the algorithm, which terminates when it is not possible to grow at least one of the current set of common substructures by the inclusion of a further atom.

An alternative algorithm is available from the work which has been done on *clique* detection in graphs, where a clique is a subgraph in which every node is connected to every other node and which is not contained in any larger subgraph with this property [21]. Given a pair of 3-D chemical structures A and B, a *correspondence graph, C*, can be formed in which the nodes are pairs of atoms ($A[U]$, $B[X]$), one from each of the two structures, such that $A[U]$ and $B[X]$ are of the same atomic type. Two nodes ($A[U]$, $B[X]$), ($A[V]$, $B[Y]$) are marked as being connected by an edge in C if the values of the edges from $A[U]$ to $A[V]$ in A and from $B[X]$ to $B[Y]$ in B are the same. Maximal common subgraphs then correspond to the cliques of C, and structurally similar molecules can hence be identified by clique detection, for which there are many algorithms available: of these, the backtracking tree search algorithm due to Bron and Kerbosch [13] seems to be that which is best suited to the processing of 3-D chemical structures [10].

Brint and Willett describe an extended comparison of the efficiencies of the Crandell–Smith and clique detection algorithms using molecules from the CSD and show that the clique detection method is far more efficient in operation than the Crandell–Smith algorithm [10]. The difference in performance is noticeable even when the MCS is quite small, and becomes very large indeed in the case of pairs of molecules sharing a large common substructure, the clique detection algorithm here being up to two orders of magnitude faster in execution. The Crandell–Smith algorithm should be better at dealing with large molecules, e.g., of size 40 non-hydrogen atoms or greater, because the correspondence graph for the clique finding approach then becomes very large. However, even with such molecules, the Crandell–Smith approach cannot be used if the MCS is at all large since it then incurs unacceptable storage costs [10].

In later work, the clique detection procedure was extended from its original application, the comparison of a small number of molecules to identify putative pharmacophoric patterns, to the implementation of a 3-D nearest neighbour retrieval system in which the measure of molecular similarity is the size, i.e., the number of atoms contained in, of the MCS between a 3-D query molecule and each of the 3-D molecules in the database [12]. The simple way to implement such a nearest neighbour retrieval system is to compare the query molecule with each database molecule in turn, identifying the MCS in each case using the clique detection algorithm described above. Such an approach is clearly very time-consuming and efficiencies of operation can be obtained by use of an initial upperbound calculation. This calculation allows many of the molecules in a database to be eliminated from the clique detection algorithm, which is invoked only for those molecules that have a large number of fragments in common with the query molecule. The fragments used here are the interatomic distance fragments discussed previously that were originally developed for improving the efficiency of 3-D substructure searching [19]. The implementation of these upperbound calculations is described in detail by Brint and Willett [12]. Their results suggest that the calculations will increase the efficiency of MCS detection when the query molecule has large, heteroatom-rich substructures in common with the molecules in the database that is being searched; in this case, up to 90% of the molecules can be eliminated from the time-consuming clique detection stage. In other cases, the reduction in computation can be quite small.

3.3 Fragment-based similarity

The maximal common subgraph approach provides a very strict and rigorous definition of structural similarity, since a mutually consistent set of structural elements must be identified that is present in both of the structures that are being compared; even a single mismatching distance, for example, must result in the elimination of a structural element from the matched substructure. Alternative definitions of similarity attempt to calculate an overall global measure of similarity between pairs of structures using the idea that a pair of molecules having a large number of fragments in common are expected to have a high degree of similarity. A typical measure is the *Tanimoto coefficient*: given a pair of molecules having A and B fragments respectively, C of which are in common, then the Tanimoto coefficient is defined to be $C/(A+B-C)$. Experimental studies [32] have shown that the calculated similarities provide an efficient and effective basis for a wide range of 2-D structure matching techniques, including the ranking and clustering of substructure search output, browsing in chemical databases, and the selection of compounds for inclusion in biological screening programs. There is obvious interest in seeing whether analogous fragment-based methods are applicable to databases of 3-D molecules.

This approach has been adopted in recent work by Artymiuk *et al.* [4] which sought to determine whether structural similarity between protein structures in the PDB could be determined automatically, using the linear SSEs described previously as the basis for the calculation of interprotein similarity. This study has made use of two previous classifications of proteins, due to Levitt and Chothia [22] and to Richardson [29]. Given a target molecule belonging to a known class, the aim of the work was to determine whether its nearest neighbour proteins, i.e., those that were structurally most similar to it, were in the same class. The existing classifications thus provide an external criterion for the effectiveness of the similarity calculation.

In outline, the procedure adopted was as follows:

- A target protein was chosen from the PDB, and its class in the existing classifications noted. It was represented by a list of all of the distinct interline angles and distances.
- The set of characteristics for the target compound was then compared with the corresponding sets of characteristics for each of the structures in the PDB to identify the number in common. Specifically, if the target structure, Q, has NQ SSEs, and the protein, S, with which it is being matched, has NS SSEs, the similarity measure is calculated by comparing each of the $NQ(NQ-1)$ query angles and distances with the $NS(NS-1)$ structure angles and distances to find those in common (to within any allowed tolerance).
- The number of characteristics in common is then used to calculate the Tanimoto similarity coefficient, and the coefficient value used to update a list of the nearest neighbours for the target compound.
- The two previous steps are repeated for each protein in the PDB. The nearest neighbours are then searched for in the existing protein classifications to see whether they are in the same class division as the target protein.

Tests with 17 different target structures suggest that this approach provides a reasonably effective means of predicting the class membership of a protein in the Levitt–Chothia classification, but the predictions are less effective in the case of the Richardson classification. This is, perhaps, not very surprising in that the latter classification is based much more on the structural domains of the proteins, rather than the whole structure, which is being considered here [23].

3.4 Ranking of POSSUM output

The final type of similarity calculation we have studied to date is intermediate in character between the MCS-based and fragment-based measures; this work has been undertaken to try to relax the rather stringent and arbitrary distance constraints that need to be used in the POSSUM program which has been described previously. Specifically, the aim was to provide a ranking of the output with the hope that the best hits would be clustered together at the top of the ranking since although it is generally effective in operation, POSSUM does have two characteristics which can cause problems for certain types of search:

- The precise specification of the distance information needs to be done in different ways for different motifs and thus requires a fair degree of familiarity with the system if precise searches are to be achieved.
- No explicit account is taken of the size of the SSEs, i.e., the numbers of residues contained within them, as manifested by the length of the lines in 3-D space. It is thus possible to obtain hits which, while satisfying the information in the query statement, differ considerably from the actual motif represented by this statement.

A procedure has been developed that allows the ranking of the output from an initial, broadly defined search so that the structures at the top of the ranking are those which contain motifs that are structurally most similar to the query motif. The basic idea of the ranking algorithm is to approximate the overall shape of a motif, either in a query or in a database structure, by the distribution of interline distances between the component SSEs. Whereas the basic substructure searching algorithm in POSSUM utilizes only a single interline distance, either the midpoint distance or the distance of closes approach, the ranking procedure utilizes large numbers of distances between each pair of lines. Specifically, each linear SSE is automatically assigned a series of points located at equal distances along the major axis representing that SSE. Distances are then calculated between each distinct pair of points for each distinct pair of lines and the frequency distribution of these distances calculated. The degree of similarity between a query motif and a database motif resulting from the initial search is then measured by the extent of the agreement between the two distance distributions; this similarity is calculated by means of the χ^2 statistic or of the sum of the squared differences. If the overall shape of the secondary structure elements of a database motif is very similar to that of the query motif then the two distributions should be broadly comparable. However, the distributions will differ if one or more of the secondary structure elements in the database motif is not of comparable size to the corresponding secondary structure element in the query motif (these correspondences having been identified in the initial search so that it is known, for example, which particular β-strand in the matched database motif corresponds to a β-strand that has been specified in the query motif).

Grindley *et al.* [16] discuss the use of this algorithm on typical secondary structure motifs and demonstrate that it provides an extremely effective way of searching the PDB since it does not require the user to be able to specify very precise angular or distance constraints but ensures that the best matches to the query are those that are displayed first to the user. The only problem with the algorithm is that it is extremely demanding of computational resources, owing to the very large numbers of interpoint distances that need to be calculated. However, these calculations are very well suited for a parallel implementation and we have demonstrated that the Distributed Array Processor, a massively parallel machine, allows the implementation of the ranking procedure in a highly cost-effective manner [16].

4 CONCLUSIONS

In this paper, we have presented an overview of the work which has been carried out in Sheffield over the last few years to evaluate the utility of graph matching techniques for the processing of the 3-D molecules in the CSD and in the PDB. We believe that the results that have been obtained to date provide not only a firm basis for the use of graph-theoretic methods in a wide range of 3-D structure matching contexts but also several practical algorithms for the implementation of these methods as follows:

- Substructure searching for distance-based pharmacophoric patterns in the CSD can be carried out on a routine basis using a two-stage retrieval algorithm that is entirely analogous to that used for 2-D substructure searching, with an initial screening search being used to eliminate the great bulk of the file prior to the detailed and time-consuming geometric search. This latter search is implemented most efficiently using Ullman's subgraph isomorphism algorithm. It should be noted that a 3-D substructure searching system is likely to have a much slower speed of response than a comparable system for 2-D substructure searching, owing to the $O(N^2)$ of inter-atomic distances that need to be considered in both stages of the search.
- The implementation of substructure searching in the PDB depends on the level of structural description that is available. In the case of searches for patterns of atoms, it is possible to use a modified version of the CSD geometric searching algorithm with the Ullman algorithm being preceded by a screening mechanism based on the work of Lesk; in the case of searches for secondary structure motifs, a higher-level abstraction of the protein structure has been developed that accords more closely with the requirements of research in protein engineering. This linear structure representation has proved to be well-suited to a range of structure matching algorithms and forms the basis for our continuing studies of retrieval techniques for the structures in the PDB.
- Several methods have been developed for the calculation of 3-D structural similarity, these include fragment-based similarities, the clique detection approach to maximal common subgraph detection, and the POSSUM ranking procedure. All of them seem to be highly effective in operation; they are also, however, quite time-consuming in operation, even using the efficient algorithms that we have devised for these applications. Accordingly, although we have not discussed this work in detail in the context of the present paper, we have begun to evaluate the use of parallel computer hardware to increase the run time efficiencies of some of these algorithms [11, 16].

Our studies in the general area of 3-D chemical graph matching are continuing. Thus, we have recently completed a comparison of the utility of 2-D screens and 3-D screens for property prediction in substructural analysis studies of small molecules [25] and are currently investigating the use of the Bron–Kerbosch algorithm for MCS detection in pairs of PDB structures (with the aim of being able to recognize novel secondary structure motifs automatically, these then being used as the basis for searches in POSSUM). Again, we shall shortly be commencing a statistical analysis of the occurrence of various types of angle in the CSD, with the aim of providing angle-based screens that will complement our current distance-based screening capability and thus allow the specification of both types of information in pharmacophoric pattern searches.

ACKNOWLEDGEMENTS

We thank the British Library Research and Development Department, the Department of Education and Science, Pfizer Central Research and the Science and Engineering Research Council for funding. PJA is a Royal Society University Research Fellow and DWR is a Lister Institute Research Fellow. Substantial contributions to the work have been made by Andrew T. Brint, David Bawden, Hazel M. Davies, Jeremy D. Fisher, Susan E. Jakes, Nicola J. Watts and Terence Wilson. This paper is a contribution from the Krebs Institute for Biomolecular Research, University of Sheffield.

REFERENCES

[1] Abagyan, R.A. and Maiorov, V.N., A simple qualitative representation of polypeptide chain folds: comparison of protein tertiary structures, *Journal of Biomolecular Structure and Dynamics* 5 (1988) 1267–1279.

[2] Abola, E.E., Bernstein, F.C., Bryant, S.H., Koetzle, T.F. and Weng, J., Protein Data Bank, in: Allen, F.H. *et al.*, (eds.), *Crystallographic Databases: Information Content, Software Systems, Scientific Applications* (Data Commission of the Inter-National Union of Crystallography, Cambridge, 1987).

[3] Allen, F.H. *et al.*, The Cambridge Crystallographic Data Centre: computer-based search, retrieval, analysis and display of information, *Acta Crystallographica* **B35** (1979) 2331–2339.

[4] Artymiuk, P.J., Rice, D.W., Mitchell, E.M. and Willett, P., Searching techniques for databases of protein secondary structures, *Journal of Information Science* **15** (1989) 287–298.

[5] Artymiuk, P.J., Rice, D.W., Mitchell, E.M. and Willett, P., Striking tertiary structural resemblance between the families of bacterial signal transduction proteins and of guanosine nucleotide binding proteins revealed by graph theoretical techniques. In preparation.

[6] Ash, J.E., Chubb, P., Ward, S.E., Welford, S.M. and Willett, P., *Communication, Storage and Retrieval of Chemical Information* (Ellis Horwood, Chichester, 1985).

[7] Bernstein, F.C. *et al.*, The Protein Data Bank: a computer-based archival file for macromolecular structures, *Journal of Molecular Biology* **112** (1977) 535–542.

[8] Brint, A.T., Davies, H.M., Mitchell, E.M. and Willett, P., Rapid geometric searching in protein structures, *Journal of Molecular Graphics* 7 (1989) 48–53.

[9] Brint, A.T. and Willett, P., Pharmacophoric pattern matching in files of 3-D chemical structures: comparison of geometric searching algorithms. *Journal of Molecular Graphics* **5** (1987) 49–56.

[10] Brint, A.T. and Willett, P., Algorithms for the identification of three-dimensional maximal common substructures. *Journal of Chemical Information and Computer Sciences* **27** (1987) 152–158.

[11] Brint, A.T. and Willett, P., Identifying 3-D maximal common substructures using transputer networks. *Journal of Molecular Graphics* **5** (1987) 200–207.

[12] Brint, A.T. and Willett, P., Upperbound procedures for the identification of similar three-dimensional chemical structures. *Journal of Computer-Aided Molecular Design* 2 (1988) 311–320.

[13] Bron, C. and Kerbosch, J., Algorithm 457. Finding all cliques of an undirected graph, *Communications of the ACM* **16** (1973) 575–577.

[14] Crandell, C.W. and Smith, D.H., Computer-assisted examination of compounds for common three-dimensional substructures, *Journal of Chemical Information and Computer Sciences* **23** (1983) 186–197.

[15] Downs, G.M., Lynch, M.F., Manson, G.A., Willett, P. and Wilson, G.A., Transputer implementations of chemical substructure searching algorithms, *Tetrahedron Computer Methodology* **1** (1988) 207–217.

[16] Grindley, H.M., Artymiuk, P.J., Mitchell, E.M., Rice, D.W., Willett, P. and Wilson, T., Ranking of secondary structure motifs in searches of the Protein Data Bank, In preparation.

[17] Gund, P., Three-dimensional pharmacophoric pattern searching, *Progress in Molecular and Subcellular Biology* **5** (1977) 117–143.

[18] Jakes, S.E., Watts, N., Willett, P., Bawden, D. and Fisher, J.D., Pharmacophoric pattern matching in files of 3-D chemical structures: evaluation of search performance. *Journal of Molecular Graphics* **5** (1987) 41–48.

[19] Jakes, S.E. and Willett, P., Pharmacophoric pattern matching in files of 3-D chemical structures: selection of inter-atomic distance screens. *Journal of Molecular Graphics* **4** (1986) 12–20.

[20] Lesk, A.M., Detection of 3-D patterns of atoms in chemical structures. *Communications of the ACM* **22** (1979) 219–224.

[21] Levi, G., A note on the derivation of maximal common subgraphs of two directed or undirected graphs, *Calcolo* **9** (1972) 341–352.

[22] Levitt, M. and Chothia, C., Structural patterns in globular proteins, *Nature* **261** (1976) 552–557.

[23] Mitchell, E., Willett, P., Artymiuk, P. and Rice, D., *Three-Dimemsional Substructure Searching in the Protein Data Bank* (British Library Research and Development Department Report no. 5984, London, 1988).

[24] Mitchell, E.M., Artymiuk, P.J., Rice, D.W. and Willett, P., Use of techniques derived from graph theory to compare secondary structure motifs in proteins, *Journal of Molecular Biology*, in press.

[25] Ormerod, A., Willett, P. and Bawden, D., Further comparative studies of fragment weighting schemes for substructural analysis. Submitted for publication.

[26] Read, R.C. and Corneil, D.G., The graph isomorphism disease, *Journal of Graph Theory* **1** (1977) 339–363.

[27] Richards, F.M. and Kundrot, C.E., Identification of structural motifs from protein coordinate data: secondary structure and first-level supersecondary structure, *Proteins: Structure, Function, and Genetics* **3** (1988) 71–84.

[28] Richardson, J.S., β-sheet topology and the relatedness of proteins, *Nature* **268** (1977) 495–500.

[29] Richards, J.S., The anatomy and taxonomy of protein structure, *Advances in Protein Chemistry* **34** (1981) 167–339.

[30] Tarjan, R.E., Graph algorithms in chemical computation, *ACS Symposium Series* **46** (1977) 1–20.

[31] Ullman, J.R., An algorithm for subgraph isomorphism, *Journal of the ACM* **16** (1976) 31–42.

[32] Willett, P., *Similarity and Clustering in Chemical Information Systems* (Research Studies Press, Letchworth, 1987).

The Cambridge Structural Database System

Eleanor M. Mitchell, Frank H. Allen, Olga Kennard, Cambridge Crystallographic Data Centre, University Chemical Laboratory, Lensfield Road, Cambridge CB2 1EW, UK.

The Cambridge Structural Database (CSD) was started in 1965 with three major aims in mind [1–5]:

1. The compilation of a computerized numeric database of organic, organometallic and metal-complex crystal structures determined by X-ray and neutron diffraction. The database was to be critically evaluated and made widely available to the scientific community;
2. The development of software for the search, retrieval, analysis and display of the information contained in the CSD;
3. Fundamental research utilizing the CSD and software, developed both in Cambridge and elsewhere.

Developments in the last five years have done much to fulfill these aims and to make the CSD System the fully integrated storage, retrieval and analysis package that it is today.

The involvement of a variety of different subject specialists has been an important factor in the development of the CSD System. Originally, the staff consisted of crystallographers, who participated in all tasks from abstracting to structural research, helped only by a few clerical assistants. Over the years, with the increasing size and complexity of the database and software, the number of people employed to work on the CSD System has risen to 15 full-time and nine part-time staff. They comprise skilled technical editors with degrees in chemistry, computer specialists, information scientists and several highly experienced crystallographers, who have been members of the Cambridge Crystallographic Data Centre (CCDC) for many years.

The work of this staff falls into two sections, the Database Building Group and the Research and Development Group, although some staff are involved in both areas. Both groups are headed by a principal scientist directly responsible for the work of that group. These principal scientists work very closely with the scientific director, Olga Kennard.

The CSD System is currently distributed to around 100 companies, mainly in Europe, the United States and Japan, to 31 National Affiliated Centres worldwide and around 11 academic institutions in the UK. The National Affiliated Centres are usually located in a crystallographic laboratory and are responsible for distributing the CSD System to those interested academics in their country either by supplying tape copies of the System, or by mounting the System on a national academic network.

The functions of the CCDC are shown in Fig. 1 and these are discussed in greater detail in the sections below.

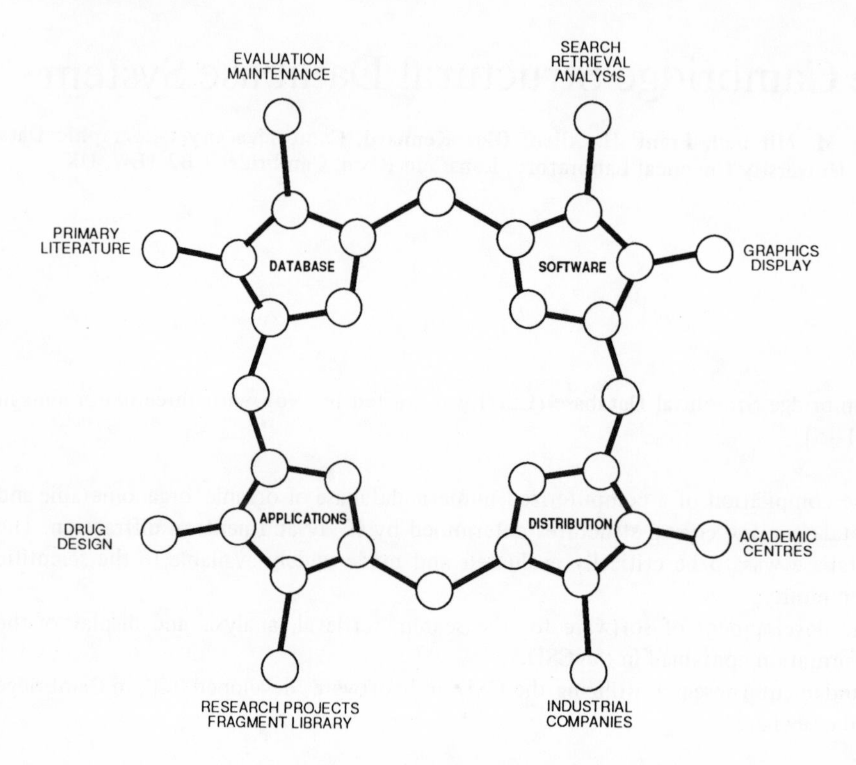

Fig. 1 – Functions of the CCDC.

THE DATABASE

The CSD covers all available results from individual crystal structure determination of organics, metallo-organics and metal complex compounds studied by X-ray and neutron diffraction. Database entries are either abstracted from the published literature or directly deposited by authors. Much of the numeric data is deposited with the CCDC by journal editors and is not published in the original paper. All entries are checked as far as possible. The checks range from the chemical name to a comparison of recalculated and published bond lengths to ensure accuracy of atomic positions. The specialist information in the original paper is transformed to make it readily usable by chemists and others not familiar with crystallography. As an example, all atomic coordinates refer to the crystal chemical unit (a covalently bonded residue) and not to the crystallographic asymmetric unit, which is a structural motif. Each entry carries a unique 6–8 letter code, originally an acronym of the chemical name but now computer generated. This so called 'refcode' is widely used as a reference when research projects involving the CSD are reported.

Fig. 2 gives an overview of the CSD System and summarizes the information content of the Database.

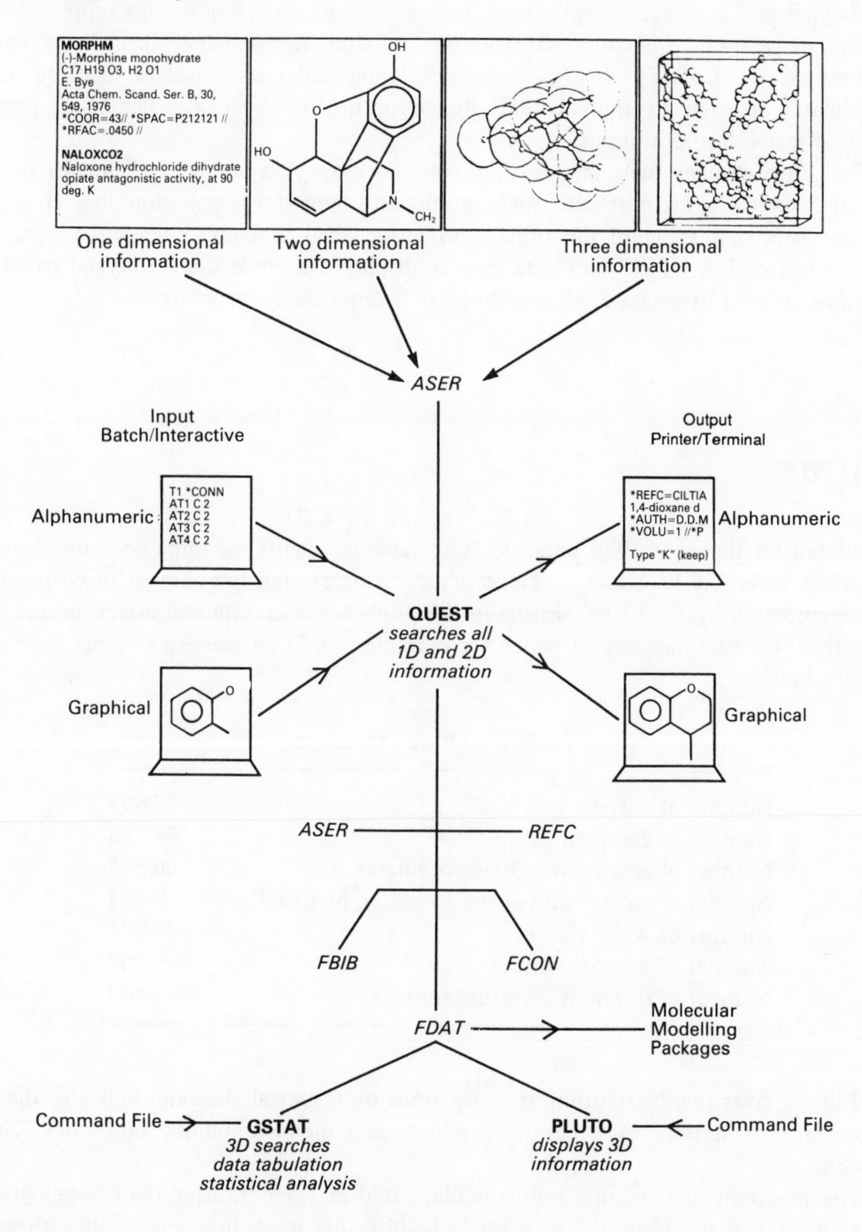

Fig. 2 – The CSD System.

The 'one-dimensional' information comprises the usual bibliographic items including chemical name, molecular formula, literature citation, chemical classification (by chemical structure type) and textual comments, i.e., neutron determination, antibiotics, etc.

The 'two-dimensional' information is the description of the chemical structure in terms of atom properties (i.e., element types, number of non-hydrogen connections, number of terminal hydrogens, net charge) and bond properties (i.e., one of eight possible bond types and a cyclicity flag). These form a compact connection table representation which can be used in chemical substructure and similarity searches. Specially extended versions of the CSD System (from Version 4 onwards) also contain digitized *x, y*-coordinates of atom positions so that publication quality chemical diagrams are part of standard search output options.

The 'three-dimensional' information comprises *x, y, z*-atomic coordinates of the crystal chemical unit; unit cell, space group and symmetry; precision indicators (R-factors, estimated standard dinations); evaluation flags and text comments; crystallographic connection tables. This data may be displayed as molecular or crystal structure diagrams, or used to generate intramolecular or intermolecular geometries.

STATISTICS

Tables 1, 2 and 3 summarize the contents of the CSD. The statistics in Table 1 are calculated on the July 1989 database. This table highlights the numeric content of the CSD, the extensive literature coverage of 563 sources, and the number of entries with errors corrected by the CCDC. Errors in the published literature still affect around 20% of entries, the vast majority of which are corrected by CCDC checking before entry into the database.

Table 1 – Overall statistics

Number of entries	77 692
Number of compounds	68 725
Number of entries with 3-D-coordinates	66 575
Number of entries with errors corrected by CCDC	9 825
Number of X-ray studies	76 985
Number of neutron studies	707
Number of different literature sources	563

Table 2 gives the distribution of compounds by chemical class and indicates the high percentage of heterocycles, natural products and metal-containing molecules (classes 62–86).

The precision of structural results is illustrated in Table 3, using the crystallographic measurement of precision, the R-factor (%). Note that fewer than 14% of the entries fall into the 'fair to bad' category.

Table 2 – Chemical class statistics

Classes	Generic coverage	Number of entries	% of CSD
1–12	Simple aliphatics	4487	5.8
13–23	Monocyclic hydrocarbons	4239	5.5
24–31	Polycyclic hydrocarbons	3360	4.3
32–42	Heterocycles	12713	16.4
43–59	Natural products	10371	13.3
60–61	Molecular complexes, clathrates	2008	2.6
62–70	Main group compounds	8319	10.7
71–75	Organometallics	12449	16.0
76–86	Metal complexes	19746	25.4

Table 3 – Precision of structural results

R	Precision	Number of entries	% of CSD
1–3	Exceptional	4529	5.8
3–4	Very high	11705	15.1
4–5	High	15353	19.8
5–7	Good	21888	28.2
7–9	Average	10781	13.9
9–10	Fair	3079	4.0
10–15	Poor	5923	7.6
15 and over	Bad	1493	1.9
Not reported	?	2941	3.8

The growth in the number of entries annually input into the database is depicted in Fig. 3. Note that although there are many entries from the literature published in 1988 and 1989 in the July 1989 database, the figures for these years are not complete and so are not included in the graphs. The number of new structure determinations is still increasing and around 8000 new entries are expected to be input in 1990. This growth is

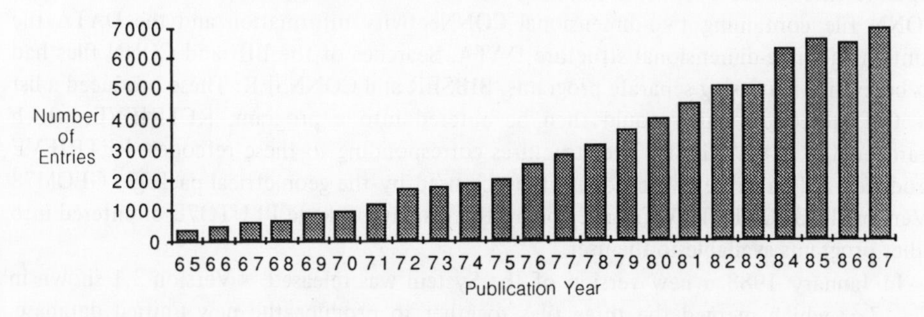

Fig. 3 – Number of entries added annually to the database.

in part the result of new technology in experimental data collection and the ever-increasing application of X-ray structure analysis. These developments in computational and experimental techniques also account for the increasing size of structures which can now be analysed (Fig. 4).

Fig. 5 is possibly the best overview of the growth of information content. Note, for example, that the amount of information processed for the 1987 entries was almost double that for 1981.

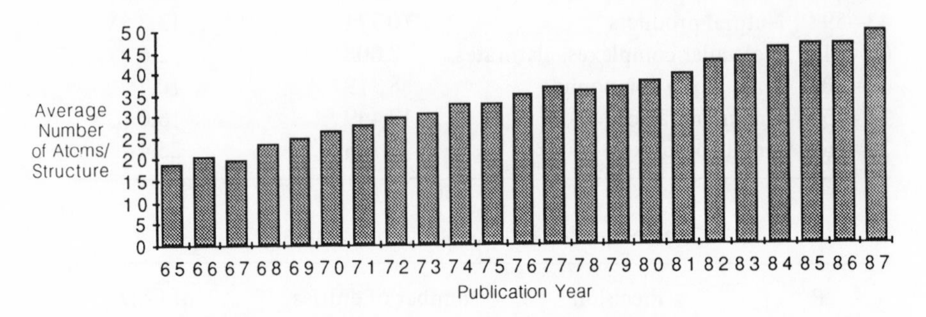

Fig. 4 – Size of the average structure deposited.

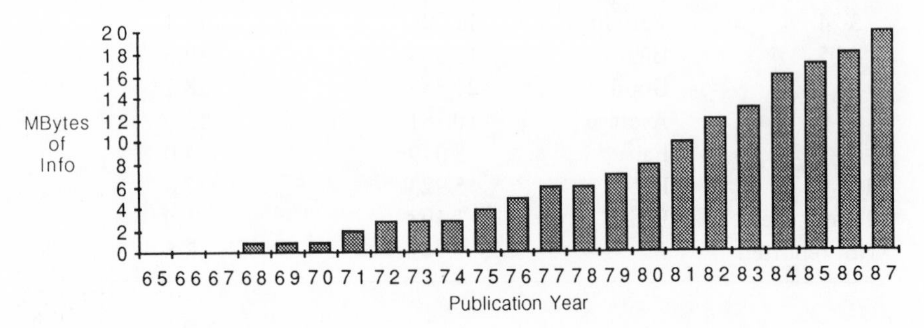

Fig. 5 – Number of megabytes added annually to the database.

EVOLUTION OF THE CSD SYSTEMS VERSIONS

The first versions of the CSD System, Versions 1 and 2 (Fig. 6), were composed of three separate files – the BIB file containing one-dimensional BIBliographic information, the CONN file containing two-dimensional CONNectivity information and the DATA file containing three-dimensional structure DATA. Searches of the BIB and CONN files had to be carried out using separate programs, BIBSER and CONNSER. These produced a list of CSD refcodes which could then be entered into a program, RETRIEVE, which searched the DATA file for the structures corresponding to these refcodes. RETRIEVE produced a data subset which could then be used by the geometrical packages GEOM78 (Version 1)/GEOSTAT (Version 2) or by the display package PLUTO78 or entered into other programs available to the user.

In January 1988 a new version of the System was released – Version 3.1 shown in Fig. 7 – which merged the three files together to produce the new unified database, called the ASER file. A search program QUEST was written to search the 1-D and 2-D

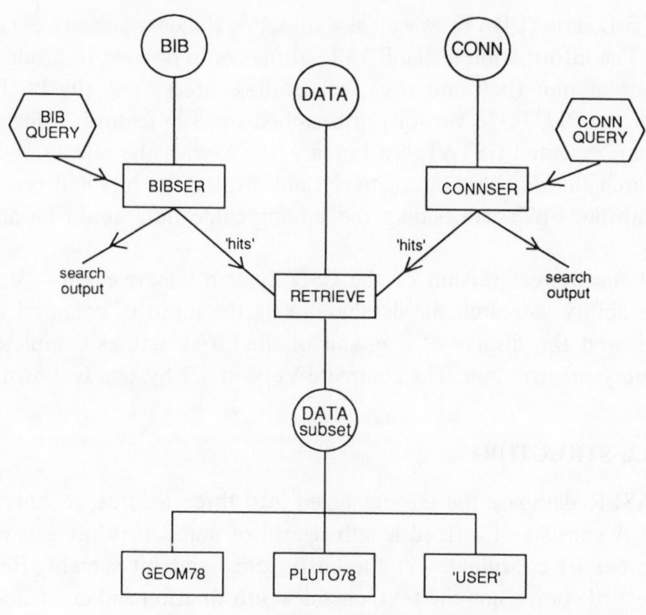

Fig. 6 – CSD System Versions 1 and 2.

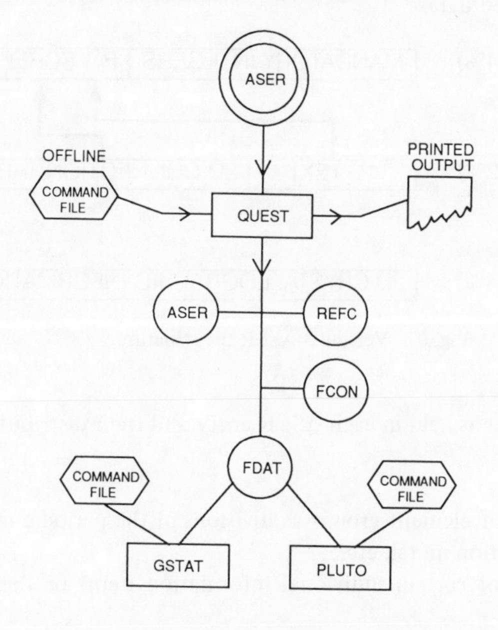

Fig. 7 – CSD System Version 3.1.

information in this file, and a set of screens was generated for the 1-D and 2-D information to speed up the search process. The search process used in QUEST is described in more detail below. This release of QUEST could only be used offline and produce printed output of searches, but could be used to produce a number of different subfiles – a subfile of refcodes (REFC file), a subfile of connectivity information (FCON) and a subfile of 3-D data (FDAT) as well as a subset of the database in CSD database file format (ASER). The information in the FDAT subfile could be used to produce a number of different types of plot (ball and stick, spacefilling, etc.) using the PLUTO package (unchanged from the PLUTO78 version) or searched for 3-D features using an enhanced geometry program, renamed GSTAT. By January 1989, with the release of Version 3.4, the ability to search the database interactively and display the hits had been added, and the remaining subfile, FBIB, containing the bibliographic data could be automatically produced.

In July 1989 the newest version of the CSD System was released – Version 4.1 – which added the ability to search the database using the input of chemical substructure diagrams and allowed the display of diagrams of the hit structures complete with highlighting of the query substructure. The complete Version 4.1 System is shown in Fig. 2.

DATABASE FILE STRUCTURE

Entries in the ASER database file are organized into three records, records A, B and C (Fig. 8). Record A consists of a fixed length record of mandatory integers (i.e., year of publication, number of coordinates in the entry, etc.) and bit screens. Record B is a variable length record comprising the text, chemical information and crystallographic unit cell data fields. Record C is also a variable length record and contains the atomic coordinates and associated data.

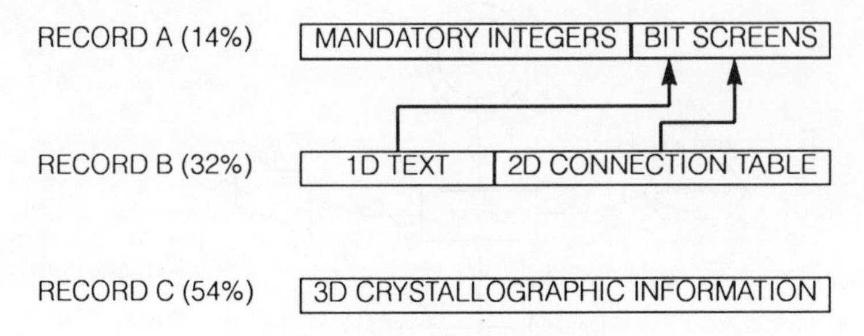

Fig. 8 – Version 3 ASER file structure.

There are 682 bit screens held in each ASER entry and their distribution falls into four categories:

(a) Presence/absence of element groups as divisions of the periodic table, e.g., halogen, alkali metal, transition metal, etc.;
(b) Presence/absence of certain individual information items or flags relating to data evaluation;
(c) Presence/absence of contiguous letter pairs in compound names and author names;
(d) Presence/absence of selected subfeatures in the chemical connection table, e.g., a triple bond, a C–P bond, a C–C=O fragment, a four-membered ring, a charged nitrogen atom, etc. The two-dimensional connectivity screens used in the CSD System were chosen after extensive statistical analysis of the entries in the database at the time of the analysis, so that they reflect the diversity of chemical structure types in the database.

Screens in categories (a), (c) and (d) are set automatically by QUEST from an analysis of the input search question command file. Screens in category (b) are set by the user if required: the user will never set screens of category (c), and will seldom set screens in categories (a) or (d).

There are 38 searchable numerical fields (excluding atomic coordinates and related material). These 38 fields may be divided into three groups: (a) commonly searched and generally useful items, (b) information relating to the crystallographic unit cell, and (c) items relating to CSD processing or to the ASER file structure, e.g.,

*ADAT : Accession date (YYMMDD) to CSD
*COOR : Number of atomic coordinates in entry
*MAXA : Maximum atomic number of elements in entry

There are 16 separate text fields in ASER. By combination or other special treatment within QUEST they give rise to 20 searchable fields, e.g.,

*AUTH : Authors' names and initials
*ERRO : Details of errors located in publication
*FORM : Molecular formula
*REFC : Reference code

There are two chemical information fields:

(a) Chemical formula, e.g., the CSD entry for 2,6-dichlorocyclohexanone dihydrate is:

C6 H8 Cl2 O1, 2(H2 O1)

This example illustrates a very important feature of the crystal structure information in the database, namely that crystal structures can contain more than one discrete bonded unit. The extra units are typically a solvent, a counter ion, or even a co-crystallized second (sometimes third) moiety of considerable size. These bonded units are termed 'residues' by the CCDC. The molecular formula of each discrete residue is listed individually within the formula field of CSD, hence searches may cover the SUM FORMULATION or (more usually) be restricted to bonded residues.

(b) Chemical connection table. This is a compact coding of the chemical structural diagram for each residue of each entry in terms of atom and bond properties.

These two different information fields give rise to five different search fields, e.g.:

*ELEM : Element and/or element groups. There are 30 preset element group symbols within QUEST based upon divisions of the periodic table; each has its own pseudoelement symbol for use in searches. Users can also define their own element groups for more specific searching.

*CONN : Search of chemical connectivity table for complete structures or substructural fragments.

The largest part of ASER, record C, stores the primary numerical results of each structure analysis. This consists of symmetry operators, covalent atomic radii, atomic coordinates and crystallographic connectivity. This material, together with the crystal data (unit cell, etc.), is used by GSTAT and PLUTO via the intermediate FDAT file generated by QUEST.

SEARCH SOFTWARE: QUEST

The search program QUEST is used to search the database for all bibliographic and 2-D structural information. The query language structure of QUEST requires that separate tests on individual information fields in the ASER file entry are defined first. A complete question may then consist of an individual test or a number of tests combined using the Boolean logical operators, AND, OR and NO.

A number of 'master' keywords are also available in QUEST (e.g. PRINT, SAVE) which control the style of printed (terminal) output for the hits, and also the subfiles of information to be created.

A very simple example of the search philosophy is:

```
SAVE FDAT
T1 *CLAS .LE. 51
T2 *MAXA .LE. 8
QUES T1 .AND. T2
```

This question will locate all entries in the organic chemical classes up to and including 51 (steroids) via T1. T2 will locate entries which contain no element with atomic number larger than 8 (oxygen). The QUEStion line will combine the results of these two tests via the .AND. operation to produce a search for entries in classes 1–51 with oxygen as 'heaviest' atom in the structure. The SAVE FDAT will instruct QUEST to create an FDAT subfile for the hits.

Use of screen settings as part of a search question can give great selectivity and control over the eventual hits obtained. The addition of two screens to the above coding would enable the user to restrict the hits obtained to be in error-free, neutron studies:

```
SAVE FDAT
SCREEN 33 49
T1 *CLAS  .LE. 51
T2 *MAXA .LE.  8
QUES T1 + T2
```

A search using SCREENs alone is also permissible. For example, the single instruction line:

```
SCREEN 57 90 48 153
```

means that entries will only be located which:

57 have organic compounds
90 have a crystallographic R-factor less than or equal to 0.05
48 have their absolute configuration established by X-ray methods
153 have atomic coordinates present

Prior to Version 4, input to QUEST was only in alphanumeric form. Definition of a chemical structure or substructural fragment could only be carried out by a user specification of the ATom and BOnd properties in a CONNectivity test, together with a number of additional keywords and sub-keywords to restrict or define the environment of the fragment. With the advent of Version 4, users can now input their connectivity searches using a graphical interface (see below). CONNectivity tests may be used as often as required in QUEST, either to locate two or more separate fragments in the same entry, or to locate fragment 1 in the absence of fragment 2 (e.g., T1 .NOT. T2).

The example below provides an illustration of the type of information required in Version 3 to define a CONNectivity test for a penicillin fragment:

```
T1  *CONN
C Penicillin derivatives
AT1   C 2
AT2   C 3
AT3   N 3
AT4   C 3
AT5   C 2
AT6   C 4
AT7   S 2
AT8   O 1
AT9   C 1 3
AT10  C 1 3
BO 1 2 1
BO 2 3 1
BO 3 4 1
BO 1 4 1
BO 3 5 1
BO 5 6 1
BO 6 7 1
BO 4 7 1
BO 2 8 2
BO 6 9 1
BO 6 10 1
END
```

The search process in QUEST is in two stages: each entry undergoes a screening stage and then a more in-depth searching stage. It is by now common knowledge that the addition of a screening stage to a search process can greatly enhance the speed of the search [6–8]. Screen matching in the CSD System is carried out by a Fortran intrinsic function which compares the bit strings for the query structure, set by the QUEST program and/or the user, and the database entry, set by the database creation program. Each screen (one bit of storage) can have settings of 0 or 1 only. A setting of 1 indicates that a pre-defined piece of information is present in the entry: a setting of 0 indicates that this information is absent. All screens set for the question MUST be present in a given ASER file entry before that entry can be considered as even a 'potential hit'. Any entries which pass the screening stage will be passed on to either a text-string or numerical comparison, or to an atom-by-atom, bond-by-bond matching stage for the chemical substructures.

VERSION 4 UPGRADE

Upgrading the CSD System to create Version 4 involved changes to both the database itself and to the associated software. The database changes in Version 4 involved a major upgrading of CSD input and check procedures. Connection tables are now derived from chemical diagrams prepared by our abstracting staff. Coordinates relating to a backlog of 30 000 connection tables, which existed at the start of the project, have been inserted. In some cases specially written computer programs generated graphical output suitable for editing, but often a complete re-digitization was required. All diagrams are closely scrutinized and edited to ensure standardization of presentation for all chemical classes. The range and complexity of the organic and organometallic structures in CSD dictated that the diagrams were entered by this process of coordinate digitization and storage rather than by one using algorithmic procedures. Record B in the ASER file was upgraded with x, y-coordinate information for all of the atoms in each 2-D connectivity table. The addition of these 2-D coordinates allows graphic illustration of the hits found in a search of the database.

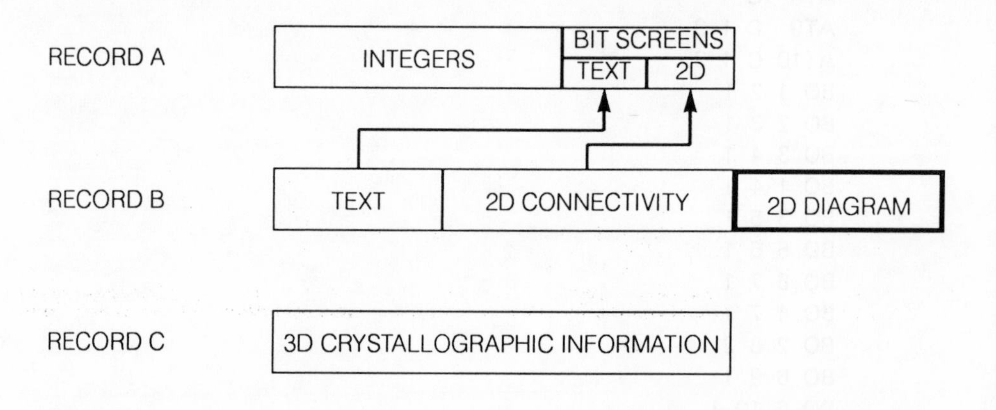

To accompany this major upgrade of the file content, a major software upgrade to QUEST was developed to permit input of substructure search queries in graphical mode. The menu-driven interface provides access to all of the powerful search features available within the Version 3 QUEST alphanumeric query language. The CSD graphics interface supports the Tektronix terminals, Tek 4010/4014, 4105, 4107, 4205 and 4207 and it can also be used on 'good' emulators of the Tektronix standard. Software requirements are an associated 32-bit computer with a Fortran 77 compiler.

There are four main menus in the System at present:

– The BUILD Menu allows chemical fragment specification, so that the alphanumeric example of the penicillin fragment above can now be input for searching into the database as:

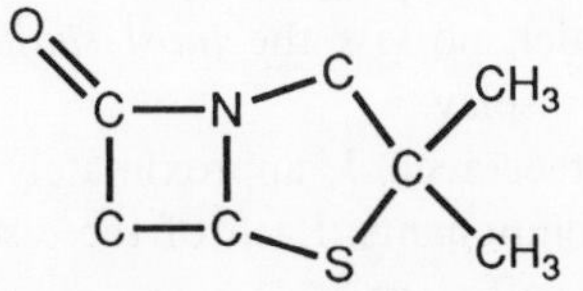

– From the Build Menu a template library of over 60 'standard' structural skeletons (e.g., hydrocarbons, steroids, etc.) and 26 pre-defined chemical functional groups (e.g., phenyl, triphenylphosphine, etc.) can be accessed.
– The EDIT Menu prcvides a powerful graphical editing facility based upon the concept of editing sentences, i.e., an entity is selected (e.g., an atom, a residue, a structure, etc.) and an action selected to be carried out on that entity (e.g., deletion, duplication, scaling, etc.).
– The SEARCH Menu allows for numeric and text queries.
– The FILES Menu provides facilities to store and retrieve search structures used in previous sessions.

There is a full, context-sensitive, on-line HELP facility and documentation is provided in an extensive user manual.

QUEST has also been upgraded with additional software for the production of chemical diagrams from the stored *x*, *y*-coordinate information. In response to a search, the query substructure(s) are highlighted on the resultant diagram for easy visual assimilation. For example, the penicillin substructure entered above produces a hit in the Ampicillin structure below:

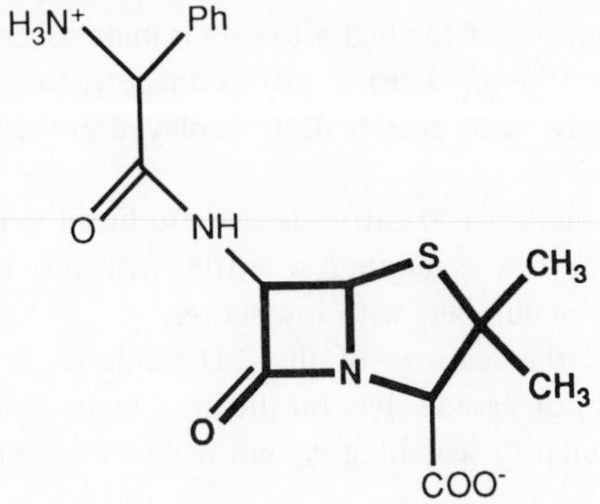

CSD System Version 4 therefore represents a completely integrated chemical graphics upgrade to the basic Version 3 System.

SIMILARITY SEARCHING

The CSD System also currently offers a 2-D similarity searching feature for browsing through the database. Similarity searching is useful in the retrieval of a structure and all its near neighbours, e.g., a compound and its derivatives (extra methyls, etc.) or an acid or base and its salts. The feature is particularly useful if a substructure search for a particular fragment has been carried out which produced no hits in the database. Similarity searching allows the structural features of the fragment to be used to rank all the database entries to find which possess the *most similar* structural features to the fragment and then to allow their display.

Of the 682 screens in the screen record, approximately 500 are used to record the presence of chemical structure components. Each of these screens is a 'feature' which is used in calculating the similarity coefficient.

Similarity between a pair of objects (x, y) to which binary descriptors (screens) have been assigned may be measured by calculating four numbers:

A the number of features in common
B the number of features unique to 'x'
C the number of features unique to 'y'
D the number of features not in 'x, or 'y'.

There are several 'similarity coefficient' formulae, for a review see [9]. For chemical similarity, features which are mutually absent (D) are not very helpful, therefore formulae which use this figure can be discounted (such as the Simple coefficient). Several studies have shown that the Tanimoto (Jaccard) or Dice coefficients are an efficient measure (e.g., [10]). The ranking produced by these two coefficients is almost always exactly the same, and so the simpler Tanimoto coefficient is the default in the CSD System; however, the user can opt for the Dice coefficient if preferred.

The similarity search fragment is defined either using Version 4 graphic input or ATom and BOnd records in a manner analogous to a CONNectivity search. The program analyses the fragment and determines its features (screens set). The screen records of all the entries in the database in turn are examined and a similarity coefficient calculated for each of them – a list of the refcodes with the highest score is maintained. At the end of the pass through the database, the 100 top scoring entries are retrieved and put into the ASER database subset format ready to be searched, or displayed in their ranked order using the graphic output.

A distribution table of the top 100 entries is also produced to give an indication of the variety of scores. This comprises usually a few entries with very high scores, more entries with medium scores and large numbers with low scores.

It is planned to extend the features of the 2-D similarity search system so that the calculation of similarity is not based solely on the bit screens assigned to the structure. It is also hoped that a 3-D similarity searching system will be available in the near future.

GSTAT

GSTAT is the main program used to analyse the numerical, experimental data held in the database. Its functions are summarized below:

(a) Entry-by-entry functions

Calculation of 'standard' molecular geometry (intramolecular, intermolecular coordination sphere)
Output of atomic coordinates in various forms (fractional, orthogonal, molecular axes)
Output of line-printer plots.

(b) Systematic tabulation of 'fragment' geometry

Definition of chemical fragment
Calculation of centroids, vectors, planes
Transformation of basic parameters
Definition of table contents.

(c) Statistical/numerical analysis of fragment and fragment geometry

Histograms and scattergrams
Principal component analysis
Correlation and regression cluster analysis (from January 1990)
Output of coordinates/plots (as above)
Superposition of fragments.

(d) Selection/rejection of entries/fragments

Entry selection on R-factor, evaluation flags, etc.
Fragment selection by geometrical criteria
'Direct' fragment selection.

APPLICATIONS OF THE CSD SYSTEM

The CSD System is becoming an increasingly important tool for both fundamental and applied research. Because the database contains information on the 3-D structure of molecules, in the form of coordinates of covalently linked residues, and also on the packing of the molecules in the crystal lattice, the information can be used for systematic analyses of intramolecular geometry and of intermolecular interactions. The CSD is the only body of information by which these latter studies can be carried out.

Within the scope of this background paper it is only possible to give one or two examples of each type of application. The reader is referred to a more detailed account [3] ; an extensive list of references can be provided upon application to the CCDC.

SYSTEMATIC ANALYSIS OF INTRAMOLECULAR DATA

(a) Interatomic distances

The determination of molecular geometry is of vital importance to our understanding of chemical structure and bonding. As a first step to utilizing the CSD for the derivation of mean geometries we have recently published two sets of definitive tables of bond length. The first of these tables lists the average length for bonds involving the elements H, B, C, N, O, F, Si, P, S, Cl, As, Se, Br, Te and I [11]. In these tables mean values are presented for 686 different bond types. The bonds are classified by common functional groups, rings and ring systems, coordination spheres, etc. Great care was taken with the statistical treatment of the samples and the investigation of outliers, i.e. contributors which differed significantly from mean values. A second publication, [12], gives similar tables for average bond lengths in organometallic compounds and metal complexes.

(b) Mean geometries

The mean geometries of complete chemical residues have been determined for furanose, pyranose and nucleic acid residues [13, 14]. The mean values were used to derive orthogonal coordinates for 'standard' residues. Such computerized 'Dreiding models' are invaluable for model building, parameterization of empirical force fields and interpretation of new structural data. They are also routinely used in crystallographic research and form part of the standard computer packages for refinement of oligonucleotide structures.

(c) Analysis of ring conformations

Systematic analysis of solid state conformation is an essential adjunct to modelling studies and prerequisite to the development of a fragment library of molecular residues. Studies of a number of ring systems have been reported, some by members of the CCDC [15]. The analysis program GSTAT contains a number of statistical functions for the automatic analysis of ring conformations, many of which were developed as part of these research projects.

(d) Drug design

Although the CSD is used by over 100 pharmaceutical and chemical companies there have been relatively few publications, for understandable reasons, on drug design. An important research paper from the Department of Pharmaceutical Chemistry, University of California at San Francisco, and the Medical Research Division of the Lederle Laboratories, uses the CSD to find a variety of chemical structures that are complementary to the shape of a macromolecular receptor site whose three-dimensional structure is known from X-ray studies. Such an approach is likely to generate new ideas for rational drug design. Other approaches in this area are based on concepts of molecular similarity. Similarity searches based on 2-D chemical characteristics are now in common use. The extension of this concept is 3-D which is of considerable interest and importance in drug design.

SYSTEMATIC ANALYSIS OF INTERMOLECULAR INTERACTIONS

One of the most interesting examples for the use of the CSD for search purposes is the study of hydrogen bonding; for a comprehensive review see [16]. A CCDC research study led to the widespread acceptance of the existence of the C-H. . .O bond [17]. This bond has been found to play an important role in stabilizing certain DNA drug complexes.

The most recent study of hydrogen bonding patterns, using the CSD, [18] has recently been highlighted in *Nature* [19] in view of its potential for the design of frequency doubling crystals which are likely to have widespread application in the rapidly developing field of optoelectronics.

These few examples are illustrations of some artificial intelligence studies using the CSD System and give an indication of its potential application to a wide variety of fields.

Version 4 of the CSD System and the forthcoming Version 5 with full chemical and 3-D functionality aim to provide the full range of tools for such studies.

ACKNOWLEDGEMENTS

Many thanks to all the people who, over the years, did the work on which this paper is based. Special thanks to Clare Macrae who stepped into the breach and presented this paper in Durham at such short notice.

REFERENCES

[1] Kennard, O., Watson, D.G., Allen, F.H., Motherwell, W.D.S., Town, W.G. and Rodgers, J. (1975) Crystal clear data. *Chemistry in Britain,* **11**, 213.

[2] Allen, F.H., Bellard, S., Brice, M.D., Cartwright, B.A., Doubleday, A., Higgs, H., Hummelink, T., Hummelink-Peters, B.G., Kennard, O., Motherwell, W.D.S., Rogers, J.R. and Watson, D.G. (1979) The Cambridge Crystallographic Data Centre: computer-based search retrieval, analysis and display of information. *Acta Crystallographica* **B35**, 2331–2339.

[3] Allen, F.H., Kennard, O. and Taylor, R. (1983) Systematic analysis of structural data as a research technique in organic chemistry. *Accounts of Chemical Research* **16**, 146–153.

[4] Allen, F.H. and Kennard, O. (1987) Cambridge Structural Database: current applications and future developments. In *Crystallographic Databases: information content, software systems, scientific applications*. pp. 55–76. Data Commission of the International Union of Crystallography. Bonn/Cambridge/Chester.

[5] Bellard, S. (1987) Cambridge Structural Database: the current system. In *Crystallographic Databases: information content, software systems, scientific applications.* pp. 32–54. Data Commission of the International Union of Crystallography. Bonn/Cambridge/Chester.

[6] Crowe, J.E., Lynch, M.F. and Town, W.G. (1970) Analysis of structural characteristics of chemical compounds in a large computer-based file. 1. Non-cyclic fragments. *Journal of the Chemical Society (C)*, 990–996.

[7] Adamson, G.W. Lynch, M.F. and Town, W.G. (1971) Analysis of structural characteristics of chemical compounds in a large computer-based file. 2. Atom-centered fragments. *Journal of the Chemical Society (C)*, 3702–3706.

[8] Adamson, G.W. Lowell, J., Lynch, M.F., McLure, A.H.W., Town, W.G. and Yapp, A.M. (1973) Strategic considerations in the design of a screening system for sub-structure searches of chemical structure files. *Journal of Chemical Documentation* **13**, 153–157.

[9] Willett, P. (1987) *Similarity and Clustering in Chemical Information Systems.* Chemometrics Series No. 12. Bawden, D. (ed.) Letchworth: Research Studies Press Ltd. New York: John Wiley and Sons Inc.

[10] Willett, P. and Winterman, V. (1986) A comparison of some measures for the determination of intermolecular structural similarity: measures of inter-molecular similarity. *Quantitative Structure–Activity Relationships* **5**, 18–25.

[11] Allen, F.H., Kennard, O., Watson, D.G., Brammer, L., Orpen, A.G. and Taylor, R. (1987) Tables of bond lengths determined by X-ray and neutron diffraction. Part 1. Bond lengths in organic compounds. *Journal of the Chemical Society, Perkin Transactions* **II**, S1–19.

[12] Orpen, A.G. Brammer, L., Allen, F.H., Kennard, O., Watson, D.G. and Taylor, R. (1989) Tables of bond lengths determined by X-ray and neutron diffraction. Part 2. Organometallic compounds and co-ordination complexes of the d- and f-block metals. *Journal of the Chemical Society, Perkin Transactions*, S1–83.

[13] Taylor, R. and Kennard, O. (1982) The molecular structures of nucleosides and nucleotides. Part 1. The influence of protonation on the geometries of nucleic acid constituents. *Journal of Molecular Structure* **78**, 1–28.

[14] Taylor, R. and Kennard, O. (1982) Molecular structures of nucleosides and nucleotides. 2. Orthogonal coordinates for standard nucleic acid base residues. *Journal of the American Chemical Society* **104**, 3209–3212.

[15] Allen, F.H., Doyle, M. and Taylor, R. (1990) *Acta Crystallographica B Series*, in press.

[16] Taylor, R. and Kennard, O. (1984) Hydrogen-bond geometry in organic crystals. *Accounts of Chemical Research* **17**, 320–326.

[17] Taylor, R. and Kennard, O. (1982) Crystallographic evidence for the existence of C-H...O, C-H...N and C-H...Cl hydrogen bonds. *Journal of the American Chemical Society* **104**, 5063–5070.

[18] Etter, M.C. (1989) *Accounts of Chemical Research* submitted.

[19] Harris, K.D.M. and Hollingsworth, M.D. (1989) Losing symmetry by design. *Nature* News and Views **341**, 19.

Reactions and synthesis

Computer treatment of chemical reactions and synthetic problems

Glen A. Hopkinson, Tony P. Cook, and Ian P. Buchan, ORAC Ltd., 175 Woodhouse Lane, Leeds, West Yorkshire.

INTRODUCTION

Reaction information systems, such as ORAC and REACCS, have many features in common with information systems designed to handle chemical structures, such as OSAC and MACCS. Reaction-indexing systems need to represent and search chemical structures and related data. In addition, they must handle chemical reactions which pose unique challenges, most notably the representation of structural change occurring over time under specific conditions.

In designing ORAC and OSAC we endeavoured to ensure that the user interface for the two systems were consistent and intuitive, minimizing the need for training and making it easy to use the two systems in parallel. Integration is also a key issue, allowing data to be transferred between different systems and presented to the end user in a variety of report formats. However, in order to allow an organization to capitalize on the ever increasing body of chemical knowledge stored in database form, a reaction-indexing system must first and foremost be able to answer queries posed by the user quickly and accurately.

REACTION INDEXING

The first generation of reaction-indexing programs developed in the early 1980s utilized existing structure-handling database techniques to represent reactions. Standard connection table datastructures were used to store reactant and product compounds. This representation of reactions supported the following search options:

- *Single exact search*. Individual structures present in a reaction could be located and distinguished as either reactant or product.

- *Collective exact search.* An overall reaction step could be identified by the presence of specific reactants and products.
- *Substructure search.* Fragments of individual structures could be located and qualified as being present in the reactant, product or both. Reaction changes could be identified by the appearance or disappearance of substructure fragments. However, this indexing mechanism was prone to mis-classify reactions if the number of occurences of a fragment in a structure was not taken into account and suffered from other specificity problems inherent in such fragment-based approaches.

The second generation of reaction-indexing systems extended the representation of reactions to include mapping information, i.e. the correspondence of reactant and product atoms and bonds involved in the reaction (Fig. 1). Since its inception, the ORAC program has had the facility to automatically identify the atom-to-atom correspondences between reactants and products. ORAC uses a method based on identifying the maximum common subgraph [1] between reactant and product structures to identify the reacting atoms and bonds. Mapping the atoms and bonds of reactants going to products provides the following information:

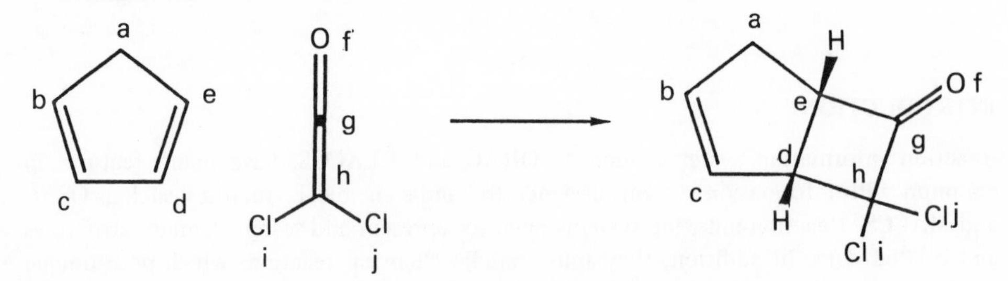

Fig. 1 – Mapping of reactant and product atoms.

- *Bond order changes.* Corresponding (mapped) bonds are found and hence the changes in bond order with respect to the other compound are added as bond labels to the individual structure graphs.
- *Stereo changes.* Stereo centres that are coincident in the reaction map can be compared and classified as Retained, Inverted, Formed or Destroyed. These atom or bond labels can then be added to the individual structure graphs.

The additional atoms and bond labels derived from the reaction mapping improve the specificity available to the substructure search. The user is able to define bond order changes and stereo changes when defining a substructure query. This extension to substructure search, while giving excellent results, puts the emphasis on the user to analyse and define which structural changes are occurring in the reaction (and which are not). In addition, it is not possible in many systems to specify the atom and bond correspondences between reactant and product structures in order to make the search more specific.

These problems were overcome in ORAC by the development of the REACTION search option, the reaction-indexing analogue of substructure search. The REACTION search option allows the user to draw in the reactant and product substructures involved in the reaction of interest (Fig. 2). The query may be made more specific by the inclusion

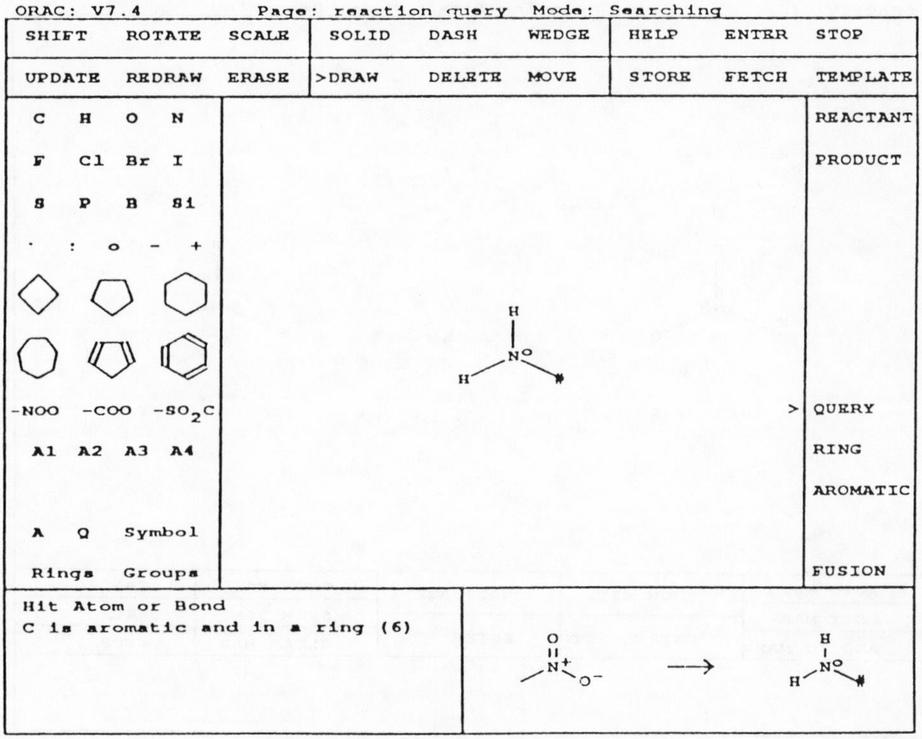

ORAC: V7.4 Page: reaction query Mode: Searching

SHIFT ROTATE SCALE SOLID DASH WEDGE HELP ENTER STOP

UPDATE REDRAW ERASE >DRAW DELETE MOVE STORE FETCH TEMPLATE

C H O N

F Cl Br I

S P B Si

-NOO -COO $-SO_2C$

A1 A2 A3 A4

A Q Symbol

Rings Groups

REACTANT

PRODUCT

> QUERY

RING

AROMATIC

FUSION

Hit Atom or Bond

C is aromatic and in a ring (6)

Fig. 2 – The REACTION Sketchpad in ORAC.

of substructure features unchanged by the reaction but still of importance in locating relevant chemistry. Once the substructures have been defined, the atom and bond mapping is performed either manually or automatically by the program (Fig. 3). From the mapping information, the program identifies the atom and bond changes occurring in the reaction and performs the required search of the reaction database.

REACTION GRAPHS

Over the last few years, a chemical information vocabulary has been developed for describing and classifying reactions [2, 6]. A *reaction graph* (RG) is a structure graph augmented with labels for breaking bonds in the reactants and forming bonds in the product. An example RG is shown in Fig. 4. The bonds in the reactant and product structures are augmented with symbols to indicate the bond order changes occurring in the reaction. The symbols used are shown in Fig. 5. The rows represent the initial bond order. The columns represent the final bond order. The symbols at the row/column intersection gives the overall bond order change.

The *reaction skeleton graph* (RSG) includes only the changed atoms and bonds; all non-reacting atoms and bonds are discarded. RSGs characterize the *type of reaction*, and are the basis of the classification technique used by the REACTION SITE search option in the ORAC program. Each reaction is indexed based on its RSG representation.

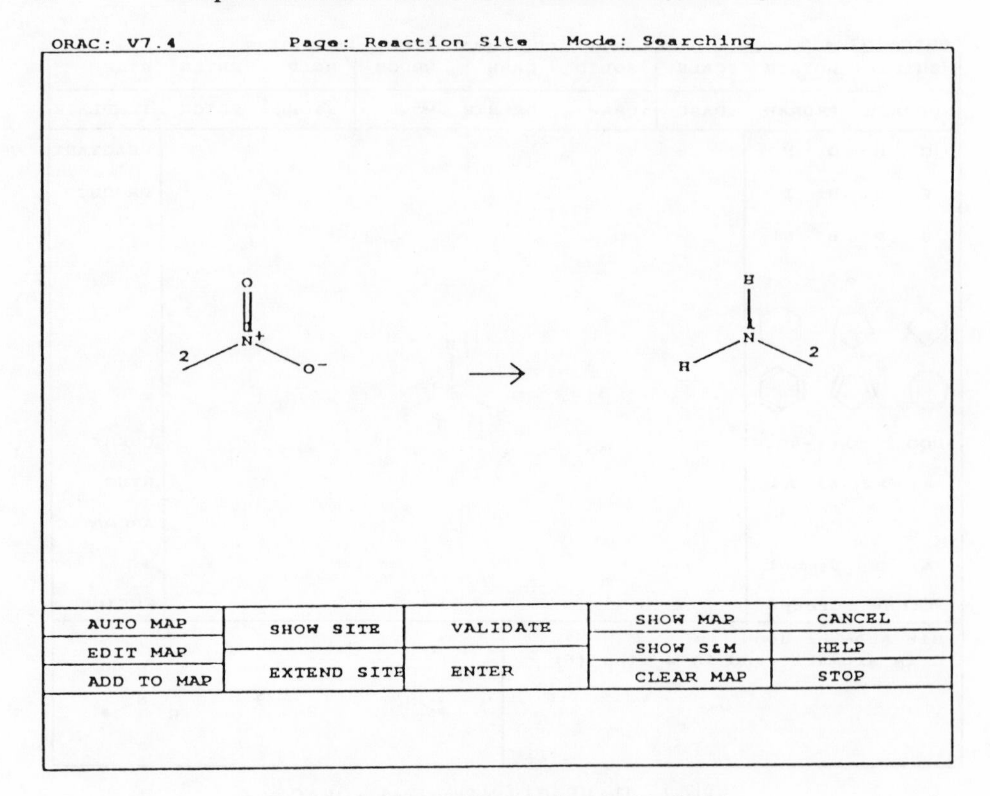

Fig. 3 – The REACTION Mapping Menu in ORAC.

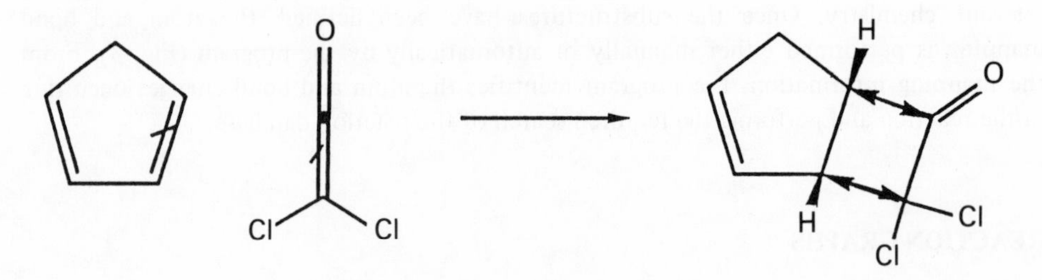

Fig. 4 – An example reaction graph.

Similarly, the query is converted to an RSG and compared with entries in the database. The RSG of the 2+2 cycloaddition reaction is shown in Fig. 6.

This reaction representation and classification approach can be extended using multigraphs. Multigraphs are formed by superimposing two graphs, in this case the RGs representing the reactant and product structures and the bond changes occurring in the reaction. Termed *superimposed reaction graphs* (SRG), the superimposition is dictated by the mapping of the atoms and bonds across the reaction. An example of an SRG scheme is shown in Fig. 7. The SRG has the following advantages over the RG representation in terms of implementation in a reaction-indexing system:

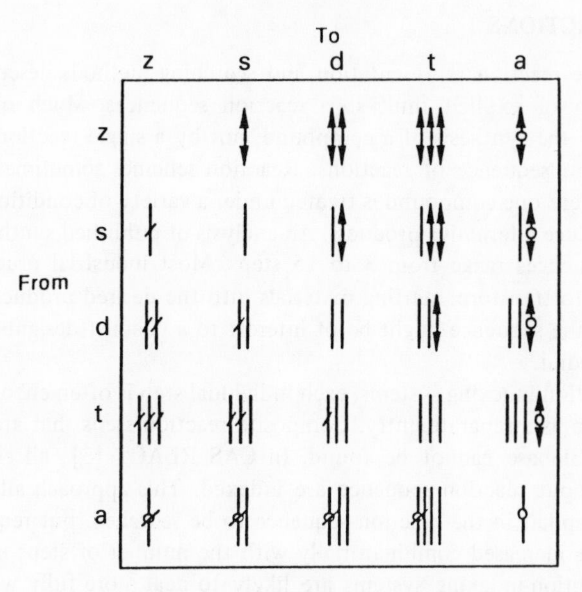

Fig. 5 – Reaction graph bond labels.

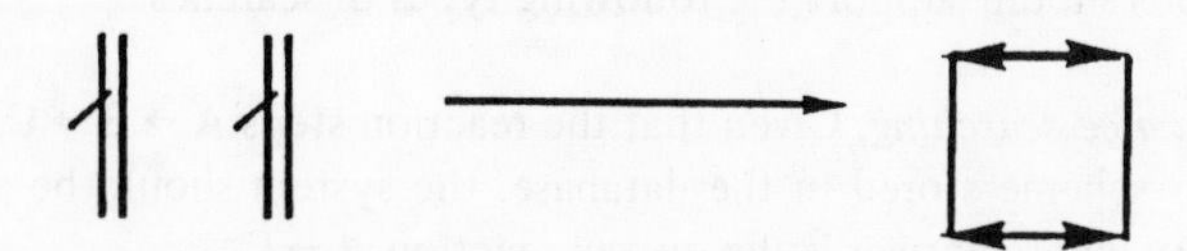

Fig. 6 – Reaction Skeletal graph for 2+2 cycloaddition.

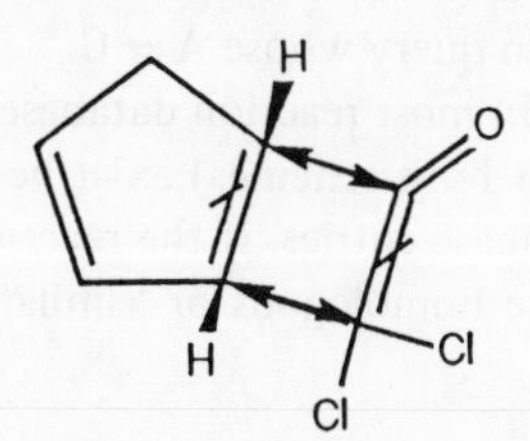

Fig. 7 – A superimposed reaction graph.

- It is more compact than the equivalent RG form. Atoms that appear in both reactant and product are represented only once.
- The mapping of reactant atom to product atom is implicit. The RG form needs to be augmented by a reaction map to complete the full reaction description.
- A variety of RGs may reduce to one SRG representation. This is true when non-stoichiometric reactions are represented. That is a reactant or (by-)product may have been omitted from the reaction scheme.
- The reactant and product structure graphs are readily derived from the SRG. Removing the arrowed bonds and converting the crossed bonds to normal bonds gives the reactant graphs; removing the crossed bonds and converting the arrowed bonds to normal bonds gives the products.

MULTI-STEP REACTIONS

None of the above reaction representation and searching methods described set out to solve the problem of explicit multi-step reaction sequences. Much of the chemical literature describes the synthesis of a compound, not by a single reaction step, but by a linear or convergent sequence of reactions. Reaction schemes sometimes also represent divergent steps where one compound is treated under a variety of conditions with a range of reagents to produce alternative products. An analysis of published syntheses reveal that most reaction sequences range from 3 to 15 steps. Most industrial processes involve a sequence of steps to transform starting materials into the desired product. Quite often a particular part of the sequence might be of interest to a chemist designing the synthesis of a related compound.

In existing reaction-indexing systems, each individual step is often encoded, along with the overall reaction as a separate entry. Composite reaction steps that are not explicitly encoded in the database cannot be found. In CAS REACT [3], all combinations of reactions for an input reaction sequence are indexed. This approach allows any of the transformations implicit in the reaction sequence to be searched, but requires storage of information that is increased combinatorially with the number of steps in the sequence. In the future, reaction-indexing systems are likely to deal more fully with the issue of representing and searching multi-step reaction schemes. An adequate representation of multi-step reactions should support the following types of searches:

- *Explicit sequence searching*. Given that the reaction steps $A \rightarrow B \rightarrow C$ exist in a multi-step reaction scheme stored in the database, the system should be able to handle a reaction query whose answer is the overall reaction $A \rightarrow C$.
- *Implicit sequence searching*. Given that reaction steps $A \rightarrow B$ and $B \rightarrow C$ exist in different reaction schemes stored in the database, the system should similarly be able to find answers to a reaction query whose $A \rightarrow C$.
- *Novel sequence searching*. In most reaction databases, it is unlikely that implicit links (a molecule is common to both schemes) exist between different reaction in more than a few per cent of database entries. If the reaction database contained $A \rightarrow B$ and $B' \rightarrow C$, where B and B′ are homologous or 'similar', the novel scheme $A \rightarrow C$ might be of interest to the user.

TAUTOMER HANDLING

A problem common to all chemical information systems is the need to handle tautomers effectively. A structure may be input as a query in one tautomeric form, but registered in the database in an alternative form. This problem is exacerbated when the tautomerization involves reaction centres.

In the Chemical Abstracts Registry System [4], bonds in structures are defined as being tautomeric upon registration. At search time, the user must specify which bonds in the query structure are allowed to match onto such bonds. In systems, such as MACCS an REACCS, tautomers are defined as having the same molecular formula and the same skeletal connectivity; however, this definition can also include many isomers. Gasteiger [5] has described a representation for generating all structures related by bond and electron shift reactions. This approach is able to generate all possible tautomeric structures of a compound, but also many that are not true tautomers.

In the ORAC and OSAC systems, we decided to implement a tautomer identification system that more closely mirrors the chemist's view of tautomers. A knowledge base of tautomer transformation rules has been developed that encodes many of the tautomeric changes that a chemical structure can undergo. Each type of tautomerization is described as a substructure change in a rule base. New tautomer rules can easily be added to the system and existing ones modified as needed without change to the main ORAC or OSAC programs. The rule language is sufficiently powerful to allow the immediate environment of the potential tautomerization to be considered. For example, keto–enol tautomerization is only allowed when the enol form is stabilized by an adjacent group. The approach adopted in OSAC and ORAC ensures that only chemically sensible tautomers are considered when processing a user's query.

When performing an exact structure search, the user indicates whether tautomers are to be considered by use of the TAUTOMER options on the sketchpad (Fig. 8). When active, tautomer transform rules are applied to identify structures in the database that are tautomers of the query structure. The system is able to identify matches which involve tautomeric changes in more than one part of the molecule. Successful matches involving tautomers are then displayed to the user (Fig. 9). We are currently extending ORAC to identify tautomers during reaction search using the same rules as used for structure searching. This capability will need to be able to reason about tautomer changes involving

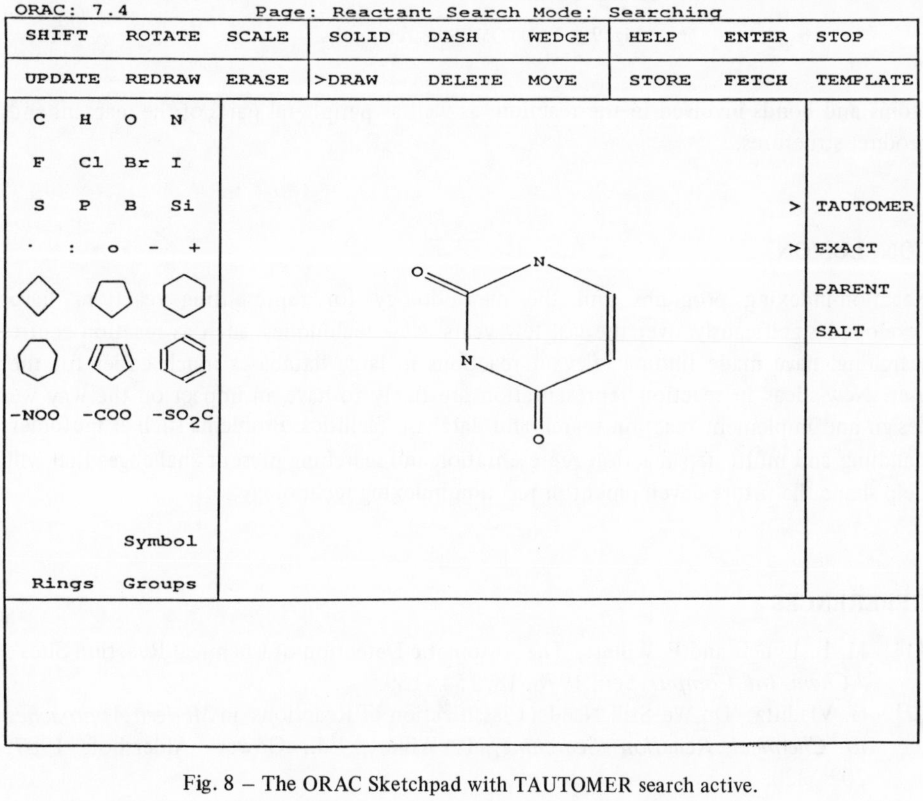

Fig. 8 – The ORAC Sketchpad with TAUTOMER search active.

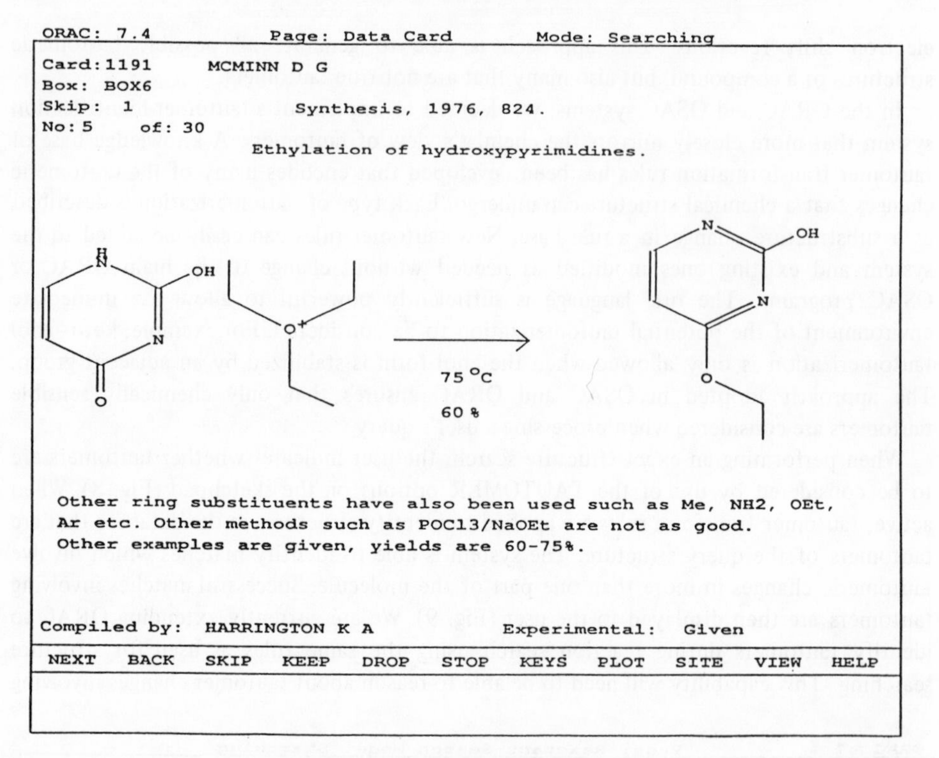

Fig. 9 – TAUTOMER search results.

atoms and bonds involved in the reaction, as well as peripheral parts of the reactant and product structures.

CONCLUSION

Reaction-indexing programs and the methodology for representing reactions have developed significantly over the last few years. New techniques, such as reaction centre searching, have made finding relevant reactions in large databases much easier for the user. New ideas in reaction representation are likely to have an impact on the way we design and implement reaction search and database facilities. Problems such as tautomer handling and multi-step reaction representation and searching present challenges that will help shape the future development of reaction-indexing technology.

REFERENCES

[1] M. F. Lynch and P. Willett, 'The Automatic Detection of Chemical Reaction Sites', *J.Chem. Inf. Comput. Sci.*, 1978, **18**, 154–159.

[2] G. Vladutz, 'Do We Still Need a Classification of Reactions' in *Modern Approaches to Chemical Reaction Searching*, P. Willett, Ed., Gower: Aldershot, 1987, 202–220.

[3] P. E. Blower Jr., S. W. Chapman, R. C. Dana, H. J. Erisman and D. E. Hartzler, 'Machine Generation of Multi-Step Reactions in a Document from Single Step Input Reactions', in *Chemical Structures: The International Language of Chemistry*, W. A. Warr, Ed., Springer-Verlag: Heidelberg, 1988, 399–407.

[4] J. Mockus and R. E. Stobaugh, 'The Chemical Abstracts Registry System VII. Tautomerism and Alternating Bonds', *J. Chem. Inf. Comp. Sci.*, 1980, **20**, 18–22.

[5] J. Gasteiger, 'A Representation of Pi-Systems for Efficient Computer Manipulation', *J. Chem. Inf. Comput. Sci.*, 1979, **19**, 111–115.

[6] S. Fujita, 'Description of Organic Reactions Based on Imaginary Transition Structures. 1. Introduction of New Concepts', *J. Chem. Inf. Comput. Sci.*, 1986, **26**, 205–212.

An overview of synthesis planning software

F. Loftus, ICI Pharmaceuticals, Mereside, Alderley Park, Macclesfield, Cheshire SK10 4TG.

THE HISTORY OF SYNTHESIS PLANNING

In the 1960s, chemists like Corey [1], Ugi [2] and later Hendrickson [3] turned their attention to analysis of the intellectual process behind what had previously been the 'art' of synthesis planning. It was they who developed much of the contemporary terminology which describes the steps and building blocks in organic synthesis; words such as 'synthon', 'umpolung', 'retrosynthesis' and 'convergence'. Along with these concepts a number of different approaches to synthesis design were developed which were aimed at a more efficient and logical procedure. Chemists quite rightly surmised that with a more logical approach computers might, some day, help them to design syntheses.

Corey felt that with the enormous variety of reactions that were available and the large numbers of chemical structures which were known it should be possible to derive some general principles by which chemists might plan a synthesis. Thus he looked for units within a target which could be formed from or assembled by known or conceivable synthetic operations. The starting point of any synthesis was a readily available substance; a 'synthon'. The alternative routes to the target could then be judged on (a) their efficiency at converting the input material to end product, (b) how well the reaction steps fit together; whether they help or hinder each other. This process was summarized by Corey [1] in the following 12 steps.

1. Simplification of problem.
2. Systematic recognition of synthons.
3. Generation of equivalent and modified synthons.
4. Addition of control synthons.
5. Systematic disconnection of synthons.
6. Formulation of the possible synthetic transformations which reform the starting structure from the derived intermediate(s).

7. Repetition of 1–6 for each intermediate and sequence.
8. Generation of intermediates until the required starting point is reached.
9. Removal of inconsistences.
10. Identification of unresolved problems.
11. Repetition of items 1–10 to generate alternative schemes.
12. Assignment of merit.

Corey [1] was the first to suggest that the cyclical, iterative process shown above looked very much like a computer program.

HISTORICAL DEVELOPMENT OF COMPUTER-AIDED SYNTHESIS DESIGN

Interestingly the first 'programs' capable of generating chemical knowledge were based on work started by Cayley [4] 133 years ago. Cayley developed algorithms capable of generating structures for all the isomers of hydrocarbons with given molecular formulae.

ORGANIC CHEMICAL SIMULATION OF SYNTHESIS (OCSS)

The first computer program to take advantage of Corey's retrosynthetic strategy was developed by Corey and Wipke in 1968. This was the Organic Chemical Simulation of Synthesis (OCSS) program. They attempted to use the vast amount of chemical knowledge which was by now available, in order to develop in OCSS an ability to work logically, backwards from the target toward smaller, more readily available fragments. Later, as OCSS began to develop, it became easy to see elements of Corey's 12 steps being used within the program. In 1969 Corey and Wipke [5] published an article on the OCSS system, which was by then one of the earliest examples of an artificially intelligent computer program.

OCSS was designed to be interactive with the chemist, allowing him to input a target structure graphically. It would then generate a 'synthesis tree' (Fig. 1) which the chemist was able to explore rapidly and in a wide-ranging fashion. This was the point at which the computer began to show its enormous power to produce and evaluate the huge number of possible reactions which lead to a target.

One problem was that with relatively complex targets this approach can rapidly generate an unmanageable number of precursors. For instance, Hendrickson [6] showed that with oestrone, which contains 18 carbon atoms, 21 bonds and 4 rings, there could be 1.07×10^{11} possible ways to construct the skeleton (Fig. 2), assuming that all the fragments used were acyclic.

Thus, the further down the synthesis tree you need to go in order to arrive at a viable starting material, the more suggestions will require evaluation. Corey and Wipke therefore used heuristics to select or reject alternatives, making the chemist's task of identifying the best route more manageable.

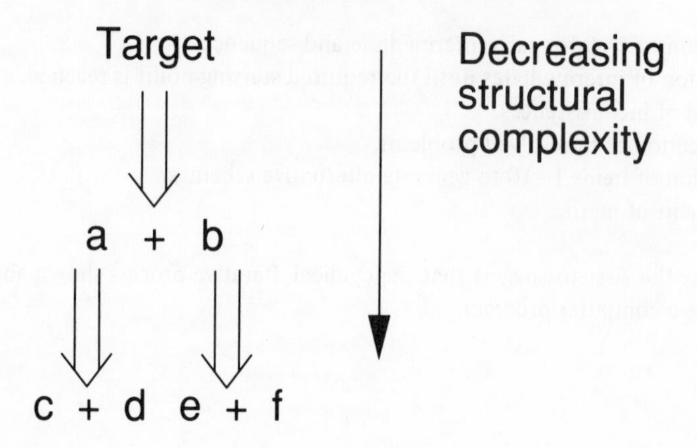

Fig. 1 – Synthesis 'tree'.

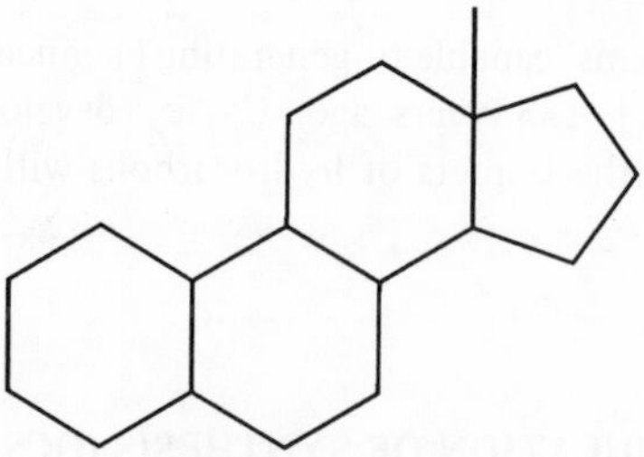

Oestrone

18 Carbon atoms
21 Bonds
4 Rings

$21!/(21-9)! = 1.07 \times 10^{11}$ ways to assemble the Oestrone skeleton

Fig. 2 – Degrees of complexity in the disconnective approach.

LOGIC AND HEURISTICS APPLIED TO SYNTHETIC ANALYSIS (LHASA)

The LHASA program (Logic and Heuristics Applied to Synthetic Analysis) [17] is the next in a series of evolutionary steps. Again, logic, which was based on retrosynthetic disconnection, provides a synthesis tree but now heuristics modifying the knowledge gained from previous searches, help to make the tree a manageable size for the chemist user.

The chemical knowledge is held in transforms (Fig. 3) which are written in a special programming language called CHEMTRN. This is very close to English; sufficiently so that chemists are able to understand the chemistry in the transform and can easily create new transforms.

```
          TRANSFORM 2400
          NAME FORMATION OF AZETIDINONES BY KETENE-AZOMETHINE CYCLOADDITIONS
          ...REF COMPREHENSIVE HETEROCYCLIC CHEMISTRY VOL.7 A.R.KATRITSKY
          ...(ED.) pp.258-260 (1984) PERGAMON
          ...REF THE CHEMISTRY OF HETEROCYCLIC COMPOUNDS, SMALL RING
          ...HETEROCYCLES PART 2. A.HASSNER (ED.) pp.223-302 (1983)
          ...INTERSCIENCE
          ...Author Frank Loftus, ICI Pharms. 5/9/85
          ...COPYRIGHT RESERVED ICI 1985
...
...
...
...
...
...
...
...
...
...
...
...STARTP
...N-C(-O)-C-C-@1
...ENDP
            RATING 15
            ....
            ADD 5 IF THERE IS A DONATING GROUP ON ATOM*5 OFFPATH
            ADD 5 IF THERE IS A WITHDRAWING GROUP ON ATOM*4
            CONDITIONS pH6:8
            ....
            BREAK BOND*1
            BREAK BOND*4
            DOUBLE BOND*3
            DOUBLE BOND*5
            ....
PATHEND
FIN
```

Fig. 3 – A LHASA transform.

Interestingly, LHASA, which is based on Corey's original concepts, is one of the longest surviving synthesis design programs and is still probably one of the most widely used in the field. This is despite there having been some 50 synthesis design programs written to date.

CICLOPS

Ugi and his co-workers [18] started to look at an algebraic approach to synthesis design, creating matrices of possible transformations. Their aim was to allow the computer to generate all the possible transformations based on fundamental principles rather than analogy with chemical knowledge. They hoped to produce syntheses which chemists had not previously thought of.

Again, a problem was the enormous number of suggestions output by the program. The generation of these suggestions therefore had to be restricted by empirical rules so that only one, two or possibly three connections were dissected at one time and only steps which proceeded at practical rates and with practical yields were considered. This latter information could be conveyed by loading areas of the transform matrix with thermodynamic data such as heats of bond formation and charge affinity information.

SIMULATION AND EVALUATION OF CHEMICAL SYNTHESIS (SECS)

This program was developed by Wipke [7] and has many similarities with LHASA. It was designed to enable the chemist to evaluate as many good syntheses as possible and, like LHASA, it works backwards from the target. Wipke terms this a type III problem (Fig. 4).

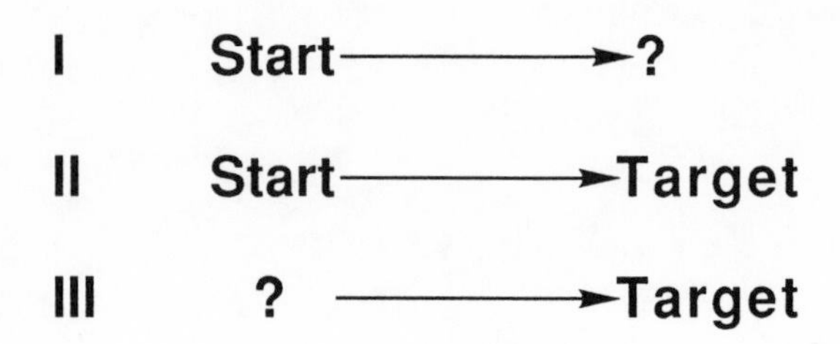

Fig. 4 – Three types of synthesis problem.

Most synthesis design programs tackle the type III problem. Wipke [8] showed that SECS could be applied equally successfully to type I and II problems.

He rightly pointed out that a thorough investigation of a reaction, leading to the synthesis of a target, T, would require a chemist to react all compounds under all reaction conditions recording the yield of T in each case. This he describes as 'Exploring the absolute space of states'. A computer can explore this space, ensuring that no good syntheses are missed, whereas a chemist might be tempted to halt a search prematurely due to the 'Eureka' factor.

SECS is transform-based with its own language, called ALCHEM, which, although similar to CHEMTRN, can also take into account the spatial, electronic and stereo-electronic effects that early versions of CHEMTRN could not.

Merck later used SECS as the basis for their Computerised Chemical Reaction Collection. The Merck chemists abstracted reactions onto pro formas and these were then used for three purposes:

1. Current awareness
2. To input into a reaction index
3. To input into SECS as transforms

SECS has also been adapted by a consortium of German and Swiss chemical companies as the CASP program.

The database of transforms has been enlarged considerably by chemists in the CASP consortium and this has now reached somewhere in the region of 6000 transforms compared with just over 2000 for LHASA. Three other facilities are available in CASP which are not in the version of LHASA used at ICI. Precursors can be 'looked up' in the Fine Chemicals Directory (hence possible starting materials can be found). Hits are sorted according to perceived merit and the complete synthesis tree can be generated and output as hardcopy.

SYNCHEM

Gelernter [9] developed the idea of looking at a target in terms of synthesis-relevant functional groups and structural features (synthemes). Each syntheme had a chapter in the reaction library which described an arbitrary number of reaction schemes for the synthesis of that particular syntheme. The program, having chosen a syntheme for further development, read the chapter into the computer and then tested the goal syntheme against each scheme. The schemes embodied the heuristics which guided the reaction. On the basis of these tests, the reaction procedure could be modified, protection specified and the merit of the proposed reaction could be adjusted if, for instance, beneficial activating groups were present. Unusually, input was by WLN rather than interactive graphical input.

Unfortunately, the program was developed in such a way that it eventually became too difficult to debug effectively. Nevertheless, SYNCHEM was used to suggest potentially useful routes. One of these is shown in Fig. 5.

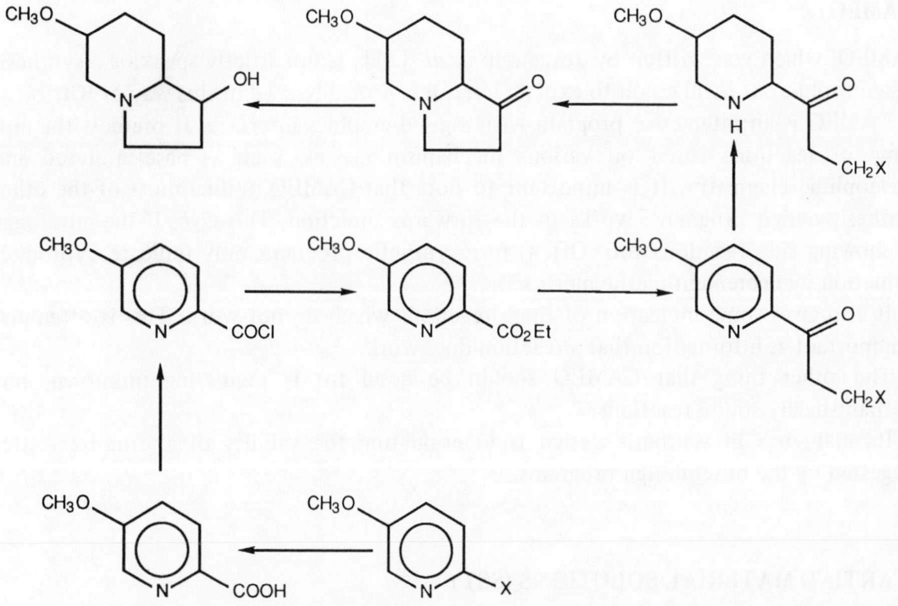

Fig. 5 – SYNCHEM – Proposed pathway to slafradiol monomethyl ether.

Work was started on a new version, SYNCHEM 2, but progress was subsequently halted when suitable computer resources were no longer available to the developers.

SYNCHEM's strengths were:

- It had a database of intermediates based on the Aldrich collection. This enabled chemists to find starting materials they had not thought of.
- The reaction compiler made entry and modification of transforms easy.
- The search algorithm took into account the 'cost' of reaching a target.
- There were heuristic devices for updating the decision tree as each cycle of sub-goals was produced.

ELABORATION OF REACTIONS FOR ORGANIC SYNTHESIS (EROS)

This program was developed by Gasteiger [19] for reaction prediction and synthesis design. It follows an algebraic approach which was stimulated by his earlier work with Ugi [10] on the CICLOPS program.

It does not rely on a database of stored reactions but creates reactions by making and breaking bonds or shifting electrons. The reactive bonds in the target are determined through calculation of the physicochemical effects for each bond in the molecule.

CICLOPS, although it was successfully used to generate a multitude of chemical reactions, also produced results of varying chemical importance. EROS seeks to remedy this by calculating thermodynamic values and thus concentrating on those reactions which are more likely to occur.

COMPUTER ASSISTED MECHANISTIC EVALUATION OF ORGANIC REACTIONS (CAMEO)

CAMEO, which was written by Jorgensen *et al.* [11], is not strictly speaking a synthesis design tool but as I will go on to explain later, it is probably used in this way at ICI.

CAMEO is an interactive program with a good graphics interface. It predicts the outcome of reactions based on various mechanism classes, such as base-catalysed and nucleophilic chemistry. It is important to note that CAMEO, unlike most of the other synthesis design programs, works in the forwards direction. This gives it the advantage of showing side reactions too. Often, retrosynthetic programs only indicate byproduct formation by decrementing the merit score.

It also gives some indication of these reactions which do not work. This is often just as important as information that a reaction does work.

The other thing that CAMEO should be good for is predicting unknown, but mechanistically sound reactions.

Its main use in synthesis design is in evaluating the validity of alternative routes suggested by the other design programs.

STARTING MATERIAL SOLUTIONS (SST)

Wipke [12] recognized that not all chemists work systematically backwards from the target. Some (most ?) take an intuitive 'leap', based on their knowledge of chemistry, back to some specific starting material.

By using SST, Wipke hoped that chemists could be assisted in the selection of a starting material. It can cope to a greater or lesser degree with the four classes of search shown below and uses a subset of the Aldrich catalogue, containing about 11 000 molecules, as its source of starting materials.

1. Target = Starting material – Exact structure search
2. Target $>$ Starting material – Superstructure search
3. Target $<$ Starting material – Substructure search
4. Target – Unlike anything

CHIRON

This interactive graphics program, was originally designed as a pedagogic tool by Hanessian *et al.* [13], to assist with teaching stereochemical perception (based on the Cahn–Ingold–Prelog rules). It has since developed into a potentially useful tool for chiral synthesis design. The search algorithm used is the Morgan algorithm. The database consists of 700 readily available chiral synthons (Chirons) culled from manufacturers' catalogues and the literature.

CHIRON shows little evidence of being influenced by other synthesis design programs (down to being written in Pascal rather than Fortran). Although it was written specifically with stereochemical synthesis in mind, it can be adapted equally well to achiral synthesis.

SYNGEN

The SYNGEN program, which was written by Hendrickson *et al* [6], was inspired by work carried out in 1976 by Velluz [14]. Hendrickson introduced the concept of convergent syntheses. His idea was to design more economic routes to targets using a convergent synthesis in which strategic parts of the target are assembled separately and then brought together later. This means that for some synthons, their participation in early, possibly wasteful stages of the synthesis is reduced. Fewer synthons are carried along at each step.

SYNGEN utilizes this principle by finding all the fully convergent bondsets in a target. It then adds functionality to the resulting skeletons in order to generate the target from a real starting material. The reactions which carry out the construction work are based on mechanistic principles.

LILITH

Written in 1987 by Sello *et al.* [15], LILITH was inspired by Hendrickson's work on convergence and aimed to find pieces of a target which were of 'equal weight'. The 'weight' involved not only the number of carbon atoms in each piece but also the number of rings and the 'complexity' of the target. Sello and his co-workers therefore produced an algorithm which could evaluate this complexity, producing precursors of similar weight and so leading to an economical synthesis.

ARTIFICIAL INTELLIGENCE FOR PLANNING AND HANDLING ORGANIC SYNTHESIS (AIPHOS)

Sasaki [16] unveiled this program in 1988. It attempts to bring together the concept of logic-based reaction generation with a knowledge-guided search strategy. This gives the chemist not only hypothetical solutions but also real reaction conditions retrieved from a stored reaction index.

USER EXPERIENCE AT ICI

ICI has made a package of four computer-aided reaction design programs available to its chemists. Until recently these programs, known collectively as SCRIBBLE, consisted of LHASA, EROS A3.2 and EROS A4.1, CAMEO and CHIRON. At present, EROS is unavailable to our users for a variety of reasons which are discussed in more detail later.

The policy for the end-user in ICI has been to allow anyone with access to the VAX-based research systems, such as ORAC, to have access to the SCRIBBLE programs also. In total, this amounts to over 600 users. Training methods vary from site to site. Some sites give day-long SCRIBBLE workshops and some teach each program to small groups. Judged by the user figures the pattern of use is nearly always the same. We see an initial burst of use following training and after this it gradually tails off to a level (based on CPU seconds) which is 10% or less of ORAC usage.

This can probably be understood if a reaction index such as ORAC is thought of as a recipe book, whereas the SCRIBBLE programs, like text books, are more fundamental. If one considers how a chemist might search for a synthesis in the library, his first stop for information about a synthesis is probably going to be something like *Organic Synthesis*, Theilheimer or maybe *Chemical Abstracts*. It is only if he does not find a suitable example in these that he will go to the text books and look for a precedent. Unfortunately this pattern of usage creates several problems when offering a synthesis design service to users such as SCRIBBLE.

Firstly, the resources required to service the various scientific programs are finite. They must be carefully managed and directed mainly at the areas which are perceived to benefit the business most. As a rough guide to usefulness, managers naturally look at the computer usage statistics. In ICI, the figures for use of the SCRIBBLE programs are low and so only a small proportion of the support resources is directed towards them.

Most of these programs are relatively difficult to learn and to operate efficiently. If the training and support resources are not there to prop up the infrequent user, they will stop using the program. This means that support resources will again be cut and so the cycle continues.

Secondly, the complexity of the programs frequently means that without expert guidance, users will at best get poor results and occasionally no useful results at all. They rapidly tire of this.

The current picture at ICI is that 30–40 research users access SCRIBBLE in any one month. Of these, roughly equal numbers access CAMEO and LHASA whereas about half that number use CHIRON. In total, there are about 1300–1400 SCRIBBLE sessions every year.

USER EXPERIENCE OF THE PROGRAMS

EROS user experience

The earlier versions of EROS used at ICI were non-interactive batch programs which ran on a remote IBM mainframe. The user interface, which ran on a VAX, was the same as that used for LHASA. This was an 'in-house' adaptation. On completion of a structure query the user was prompted to send the query to LHASA or EROS for processing. The EROS menu then prompted the user for various reaction categories. There were six of these, for example those shown in Fig. 6.

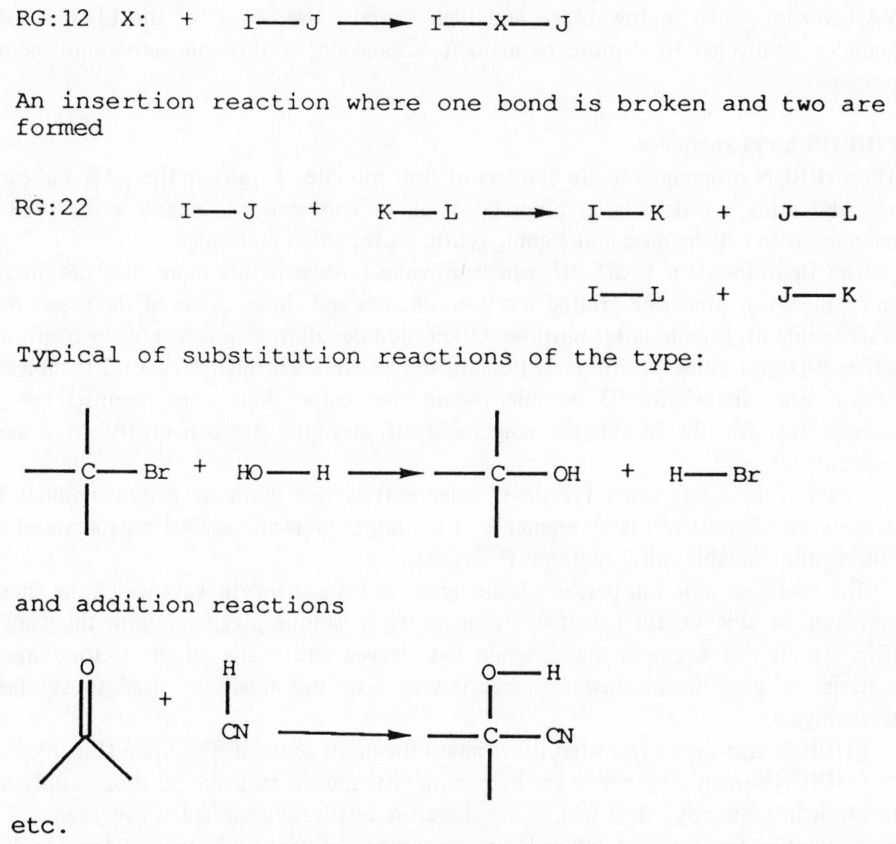

```
RG:12  X:  +   I——J  ——▶  I——X——J

An insertion reaction where one bond is broken and two are
formed

RG:22   I——J  +  K——L  ——▶  I——K  +  J——L

                              I——L  +  J——K

Typical of substitution reactions of the type:

    |                        |
  ——C——Br + HO——H  ——▶  ——C——OH  +  H——Br
    |                        |

and addition reactions

    O         H             O——H
    ‖    +    |             |
    C         CN  ——▶   ——C——CN
   / \                      |

etc.
```

Fig. 6 – EROS reaction strategies.

A simple default strategy was used by most chemists, who would add HCl as the co-product in the reaction. It was then quite easy, when looking at the output, to visualize Cl as being some other small group or atom e.g. OH, NH_2 etc. which might normally be split out from a reaction.

After the batch process had been sent, the user could return to other work on the terminal.

Although it was possible to get graphics output on the screen the best solution was to wait for the voluminous print-out from a remote line printer. This suffered from one major limitation as the printer could only draw lines in eight directions which often led to badly distorted structures. However, the advantage of hardcopy output was that it could be browsed at leisure, hopefully stimulating original ideas in the process.

The main weaknesses of EROS from the user's point of view were lack of understanding of the processes involved, the user unfriendly interface and the copious output which was produced. The last required great patience and an uncluttered mind to enable the chemist to make an intuitive leap and thus visualize a possibly novel reaction within the mass of sometimes odd-looking structures.

Later, a plan to rationalize the ICI Research computer systems onto VAX computers resulted in a temporary suspension of the EROS service. Eventually, problems with the

VAX version, such as low usage and high support overheads for the LHASA/EROS graphics input, led to a more permanent suspension of this innovative and exciting program.

CHIRON user experience

The CHIRON program actually consists of four modules. It runs on the VAX and can be accessed using a variety of graphics terminals or workstations, including the IRIS. Its main use is to aid chemists in designing syntheses for chiral molecules.

The first module is CARS-2D, which provides both structure input into the program and a means of preparing printed reaction schemes and slides. Some of the usage at ICI is undoubtedly for the latter purpose. Other modules allow stereochemical perception of either 2-D molecules, with conventionally drawn stereochemistry, or of 3-D molecules drawn using the CARS-3D module. Again, we know that some chemists use this occasionally for the automatic assignment of absolute stereochemistry to a target molecule.

CAPS (Computer Aided Precursor Selection) is the synthesis design module. This permits the chemist to match segments of his target molecule against a database of over 700 readily available chiral synthons (Chirons).

The CAPS module can perceive both target and precursors in ways which the chemist may not be able to, for example, by matching a lactone precursor onto thienamycin (Fig. 7). In this example the program has cleaved and re-shaped the lactone starting material to give the chemist a perception of how this might be used to synthesise thienamycin.

CHIRON also shows you what it has based the merit score of 87% upon (Fig. 8).

CHIRON's main weaknesses are its complexity and the fact that most users only need to use it infrequently. Most people are therefore relatively unskilled at using some of the more exciting features and will only use it for more mundane jobs such as drawing slides.

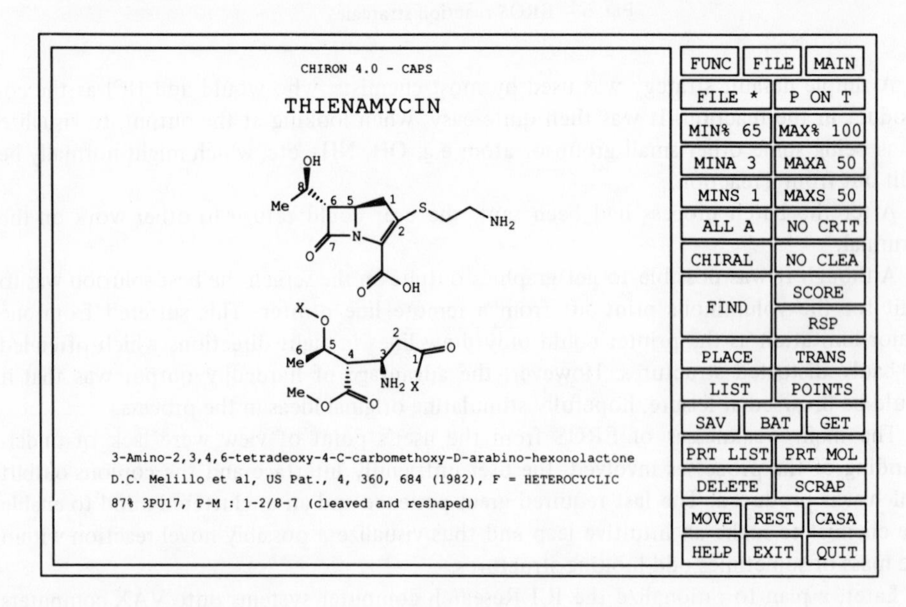

Fig. 7.

-------------------- PRECURSOR SCORING DETAILS ---------------

Name = 3-Amino-2,3,4,6-tetradeoxy-4-C-carbomethoxy-D-arabino-hexonolactone
Ref = D.C. Melillo et al, US Pat., 4, 360, 684 (1982)
File = HETEROCYCLIC (Main file) # 17
Score = 87% , cleaved between 7 and 1

Prec	Mol	MatchKind	FonctP		FonctT		Ratio	Score
1	2	Extremity	COOX		CXX		8%	50%
2	1	Chain	NoF		NoF		7%	100%
3	5	Chain	NHX	L	NHX	L	21%	100%
4	6	Branch	BRH	R	BRH	R	28%	100%
5	8	Chain	OX	R	OX	R	21%	100%
6	9	Extremity	Me		Me		7%	100%
8	7	Extremity	COOX		CONX		8%	90%

Losses and bonuses:
–4% Cleave loss
–4% Acyclic precursor wrapped target cycle

Fig. 8.

CAMEO user experience

As I have mentioned earlier, CAMEO is a program which I consider is used for synthesis design at ICI. Like CHIRON, CAMEO has several other features such as a facility for calculating pK_as. It is certainly used for this purpose by a small number of users, particularly for estimating the pK_as of carbon acids. However, its major use appears to be for testing ideas which have been generated by the chemist or by a synthesis design program. From this point of view it is not generating new chemistry; nevertheless it will frequently confirm things about a reaction that the chemist had previously only suspected.

CAMEO's major strength is that it is remarkably easy to use and occasionally it gives spectacular results. The output is immediate and understandable, and more important, very often correct.

LHASA user experience

LHASA ver. 8.2, which is currently used in ICI, permits the user to select one of three levels of expertise: advanced, intermediate or beginner. Depending on the level chosen, the user is presented with access to different options. The main strategies available remain the same but the beginner, for instance, cannot access some of the debug and test utilities.

This version is too new to get a useful view from our chemists. However, if, as I suspect it will, it leads the beginner to get a less constrained search and thus more 'noise', it will probably result in more user frustration at the output. This may be mitigated by the ease with which the output can now be generated.

Most chemists feel that LHASA is an 'experts' program. Several such experts have emerged amongst our chemists and their help and advice about using the program would normally be sought by a 'non-expert' before starting a search.

CONCLUSION

A large number of computer-aided synthesis programs are available and more are being written every year. Very often, new programs build on the experience gained from older programs but some of the older programs, like LHASA, SECS and EROS, have stood the test of time.

At ICI these programs are used by chemists but usually after other sources of information, such as reaction indexes, have been tried. The most frequent criticisms I hear are that they produce too much 'irrelevant' chemistry and that they are often difficult to use for the marginal benefits gained by the infrequent user.

As one of our users said to me the other day, 'Why use a program which at best produces the chemistry of a mediocre undergraduate when we have so many innovative minds around us? What we really need is a program equivalent to the world's best post-doctoral chemist but which can be operated by the most naive user.' Although I think we are advancing toward this goal, as a chemist myself, I can see that progress may sometimes appear frustratingly slow.

REFERENCES

[1] E.J. Corey, *Pure Appl. Chem.*, **14**, 19 (1967).
[2] I. Ugi, P. Gillespie, *Angew. Chem., Int. Ed. Engl.*, **10**, 915 (1971).
[3] J.B. Hendrickson, E.B. Keller, A.G. Toczko, *Tetrahedron, Suppl.*, **9**, 359 (1981).
[4] A. Cayley, *A. Philos. Mag.*, **13**(1), 172 (1857).
[5] E.J. Corey, W.T. Wipke, *Science*, **166**, 178 (1969).
[6] J.B. Hendrickson, D.L. Grier, A.G. Toczko, *J. Amer. Chem. Soc.*, **107**, 5228 (1985).
[7] W.T. Wipke, H. Braun, G. Smith, F. Choplin, W. Sieber, ACS Symposium Ser. – *Computer Assisted Synthesis Planning*, **61**, 97, (1977).
[8] T.M. Gund, P.v.R. Schleyer, P.H. Gund, W.T. Wipke, *J. Amer. Chem. Soc.*, **97**, 743 (1975).
[9] H.L. Gelernter, A.F. Sanders, D.L. Larsen, K.K. Agarwal, R.H. Boivie, G.A. Spritzer, J.E. Searleman, *Science*, **197**, 1041 (1977).
[10] J. Blair, J. Gasteiger, C. Gillespie, P.D. Gillespie, I. Ugi, *Tetrahedron*, **30**, 1845 (1974).
[11] T.D. Salatin, W.L. Jorgensen, *J. Org. Chem.*, **45**(11), 2043 (1980).
[12] W.T. Wipke, D. Rogers, *J. Chem. Inf. Comput. Sci.*, **24**, 71 (1984).
[13] S. Hanessian, Organic Chemistry Series Vol. 3 – *Total Synthesis of Natural Products: The Chiron Approach* Pergamon Press, Oxford, U.K. 1983.
[14] L. Velluz, J. Valls, J. Mathieu, *Angew. Chem., Int. Ed. Engl.*, **6**, 778 (1967).
[15] L. Baumer, G. Sala, G. Sello, *Tetrahedron*, **44**, 1195 (1988).
[16] K. Funatsu, S-I. Sasaki, *Tet. Comp. Method.*, **1**(1), 27 (1988).
[17] E.J. Corey, W.T. Wipke, R.D. Cramer III, W.J. Howe, *J. Amer. Chem. Soc.*, **94**(2), 431 (1972).
[18] CICLOPS – A Computer Program for the Design of Syntheses on the Basis of a Mathematical Model, J. Blair, J. Gasteiger, C. Gillespie, P.D. Gillespie, I. Ugi *Computer Representation and Manipulation of Chemical Information* Editors: W.T. Wipke, S. Heller, R. Feldman, E. Hyde, Wiley, 1974. page 129.
[19] J. Gasteiger, C. Jochum, *Topics Curr. Chem.*, **74**, 93 (1978).

Property prediction and analysis

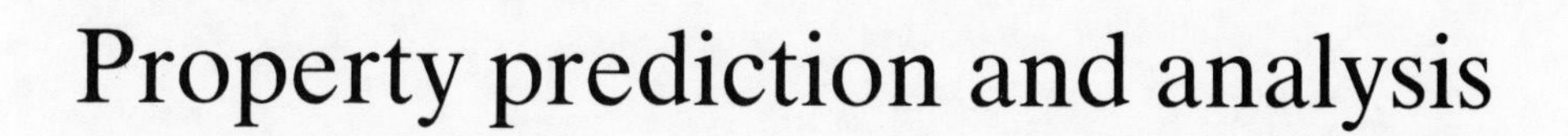

A polymorphic programming environment for the chemical, pharmaceutical and biotechnology industries

Johaina Ball, Robert V. Fishleigh, Paul Greaney, Jin Li, Alan Marsden, Eric Platt, Jennifer L. Pool, Barry Robson, Proteus Biotechnology Ltd., Marple, Cheshire.

1 INTRODUCTION

Whenever a project involves the use of several different computer programs, involves interactive computer aided design, or is combined with experimental studies, there are opportunities for ill-founded subjectivity and irreproducibility. Examples include drug design and prediction of protein three-dimensional structure. The latter is particularly topical and will be mentioned in most of the examples given. Both experimental and computational aspects of these disciplines in relation to biotechnology have been reviewed by us recently [1]. Such projects are complex and open-ended, with a great need to adapt methodologies to different circumstances and levels of availability of data. It seems unlikely that one can conceive of an objective and automatic superalgorithm in cases of this kind. We do believe, however, that this is a direction to pursue as vigorously as possible. Here we have been interested in exploring for the first time the 'super-algorithm' approach and in studying its potentialities and limitations. To this end we have taken the 'Polymorphic Programming Environment' (PPE) approach and we have developed over several years a software system for chemists and biotechnologists which incorporates operating system capabilities and a whole variety of routines for analysis, prediction and design in the molecular sciences. Most importantly for commerce, the flexibility of the system allows ways for managers to plan, interrogate and intervene, so that a whole project with financial as well as scientific considerations can be encompassed. To drive this system requires a versatile high level language with some fifth generation features that is suited to chemists. The language is called GLOBAL in so much as it draws some of its features from a variety of programming languages including PASCAL, LISP, PROLOG, and optionally some of the accepted scientific notations of FORTRAN (including formatting), with the capability of being redefined to resemble any existing or new language of choice. Further, it is particularly well suited to tackle problems which are of 'global optimization' character, i.e., with many local false solutions which trap existing programs.

2 THEORY

In considering the development of such a system it is important to consider exactly why problems such as the prediction of protein three-dimensional structure are complex and open-ended, with a need to resort to experimental data either for proteins in general or, where possible, specific to the protein of interest. As pointed out by Dirac [2]

> The underlying physical laws necessary for the mathematical theory of a large part of physics and the whole of chemistry are thus completely known, and the difficulty is only that the exact application of these laws leads to equations much too complicated to be soluble.

In other words, *everything could in theory be computed from first principles* by simulation. Why cannot computers handle the requisite equations? As stated by Boys in 1950

> It has thus been estabilished that the only factor limiting calculation of the wave function of any molecule . . . is the amount of computing necessary.

Faster computers have quantitatively, but not qualitatively, changed the situation, despite the stimulating statement 'we can calculate everything' by Clemeni in 1972. GLOBAL is designed for the new generations of supercomputers, but even the finely tuned hardware–software configuration we have developed ('The BIOENGINE') falls short of solving complex chemical problems from first principles. For example, if 100 secs are required to simulate 1 nanosecond of real time for the internal motions of a protein, and if a real protein takes of the order of a second to refold into its stable structure from an unfolded state, then 10^{11} secs would be required to predict the stable conformation by simulating the folding process. Further, the real protein molecules represent an ensemble in an effectively infinite solvent environment representing infinite degrees of freedom.

Furthermore, as part of the basic philosophy of our approach, we take the realistic view that it is not a single frozen or even vibrating structure which is to be predicted, but a pattern of behaviour embracing motion through potentially many energy minima.

We thus take the apprcach of carrying out simulations from first principles, but are guided in various ways by information other than that inherent in first principles.

In the example case of predicting protein structure, external information will be of various kinds: direct information from experiments on the protein itself (e.g. interatomic distances from nuclear magnetic resonance spectroscopy), indirect information from proteins and parts of proteins which are related to the protein of interest, indirect information about local sequence–conformation relationships in proteins in general, and subjective but rational and justifiable views of experts which can be encoded using GLOBAL. According to one view, such information other than that inherent in first principles is a limiting special case of external information which is always inherent in any simulation, since even the most objective algorithm and its use implies certain human decisions. There is a recipe (the 'Jaynes recipe') which recognizes that there is always uncertainty and that the algorithm should be constructed such that the probabilities of events arising from that uncertainty imply maximum entropy. That is, if there are several choices of probability on the basis of the available understanding, then choose those values of P which maximize $P_i \log(P_i)$. In general the use of external information can be envisaged as alternative choices of the P and to this end we frequently made explicit use of a Bayesian Information Theory approach [3]. The distribution of the P could in principle be, and in some applications actually is, progressively updated according to Bayes' theorem (posterior probability is equal to prior probability likelihood).

Probabilities in most applications can only be influenced indirectly. There are basically four approaches which we take to guiding simulations (they are actually not distinct).

(1) Use of target functions and penalty functions. For example, to model a protein fold against the folding motif of a protein domain stored on a dictionary, a function of specific atoms i, j and interatomic distance r_{ij} can be added to the interatomic force in molecular dynamics, with relative weights W and $(1 - W)$.

$$\text{EFFECTIVEFORCE}\ (i, j, r_{ij}) = W^*\text{FORCE}\ (i, j, r_{ij}) + (1 - W)^*\text{PSEUDOFORCE}\ (i, j, r_{ij}),$$

$$0 \leqslant W \leqslant 1$$

Alternatively, the function minimized in molecular mechanics can be a weighted sum of energy and penalty function

$$\text{FUNCTION}\ (C) = W^*\text{ENERGY}\ (C) + (1 - W)^*\text{PENALTY}\ (C),\ 0 \leqslant W \leqslant 1$$

where (C) is to be read as 'Function of conformation'.

Such approaches are found in practice to lead to two major classes of solution, a literal 'catastrophe' in the mathematical sense. These classes are (i) energy low, penalty function high and (ii) energy high, penalty low. For example, the situation arises when doing a combined energy minimization and root-mean squared fit of a model protein to experimental coordinates. The 'catastrophe' arises because neither the model (including energy parameters and bond geometries) nor the experimental structure is perfect, so the two terms are not *conformal*. 'Conformal' means that minima of the two functions mapped independently coincide. Note the need for weighting factor W: neglect of W implies, of course, a particular choice of $W = 0.5$. Various tricks can be done with minimization schedules in which W is varied, but in all circumstances arbitrary choices of W, or the way in which W varies, are implied.

(2) The above difficulties can to some degree be overcome by a different approach (though in purest form this is only applicable in certain circumstances). When a conformation of the molecular system is reached such that a required value of a property is achieved, but it is discovered that there are a wide variety of conformations which satisfy that property. In such a case a *conservation law* is created such that this value of the property is conserved. Natural conservation laws include conservation of total energy and momenta, such that the simulation is confined to a manifold in phase space and the manifold is one dimension less than the dimensionality of the phase space. Every conservation law created drops the accessible manifold further by one dimension. With six coordinates per atom (special coordinates x, y, z and momenta p_x, p_y, p_z conjugate to them) a large molecule has many dimensions, and many conservation laws can be imposed. For example enzyme hen egg white lyzozyme has 1001 atoms and is thus represented

in a 6006-dimensional phase space. The creation of a conservation law is in principle simple: an algorithm is written which constantly restores motion in phase space to the required value of the property. There are many approaches to this algorithm, some more efficient than others in certain circumstances, some more robust than others, some allowing considerable deviation from perfect conservation, and some allowing rapid hunting of the value of the property to which conservation is required. We are particularly grateful for an algorithm proposed by Rodney Cotterill for a slightly different application, but which is outstandingly efficient in this respect. The major advantages of this approach are (a) it avoids the use of an arbitrary weighting W, (b) it avoids some problems which arise in connection with the abovementioned 'catastrophe', and (c) it allows us to better control simulations through an understanding of the topology of manifolds in phase space. Indeed in relation to (c) it provides means, along with other tools of exploring, mapping, and understanding the properties of topologies in phase space.

(3) In many cases, it may not be meaningful to think of a bias or penalty which continuously changes with distance from a target, nor is it useful to hold to a specific value. Rather, constraints are implied outside which a simulation cannot go. In theory, these are of three types.

  (i) Explicit constraints are the values of the variables, e.g. x, y, z, coordinates or range of dihedral angles for bond rotations.
  (ii) Implicit constraints on the value of the primary function of interest, e.g. the energy.
  (iii) Implicit constraints via any other function we may wish to introduce, e.g. based on the findings of circular dichroism spectroscopy we may wish that the α-helix content of the protein being modelled lies between 10% and 30%.

True constraints of this type cannot be implemented in gradient optimization methods or force evaluation methods, since discontinuities in the derivatives are implied. The SIMPLEX approach, based on the method of [4] is one which can accommodate the above, and one of which great use is made in the modelling facilities provided by GLOBAL.

(4) Much external information cannot be envisaged as a target or penalty, or as a conservation law, because it is not readily quantifiable. Rather, the information is in the form of an 'expert opinion' associated with which there can be a degree of uncertainty. Introduction of this kind of information is best done by explicitly representing human reasoning and opinion in the computer in such a way as to select algorithms and parameters of algorithms according to results of previous steps in the protocol. These two considerations – algorithms and parameters of algorithms – are not formally distinct. In principle one should be able to express all algorithms such that they are only quantitatively different. This is actually demonstrable and indeed the GLOBAL software includes algorithms that can be become various types of molecular dynamics or energy minimization by changing the values of certain numbers and weightings in the program. Such parameters are in effect *bifurcation parameters* and there are often sudden changes in behaviour associated with shifts in the parameters.

From consideration of the above it is obvious that a high level language for our purposes be capable of a staggering variety of things including

(1) controlling and analysing 'first principle' simulations and calculations – quantum mechanics, molecular dynamics, Monte Carlo techniques, and energy minimization,
(2) developing and testing new 'first principle' algorithms,
(3) analysing and processing external information: direct experimental data, indirect data in databases, and 'expert opinion',
(4) combining the first principle simulation with the external information, and analysing the interaction,
(5) testing the quality of the results,
(6) allowing a refinement of the protocols making use of all these capabilities,
(7) providing utilities so that the mathematicians and chemical physicists can write the protocols efficiently, at a high level, and, most importantly,
(8) *allowing the means for a biologist, manager, or any person unskilled in mathematical and chemical physics principles to carry out useful computations.*

In the following we hope to convince that one can go a long way towards meeting these capabilities. It must immediately be said, however, that the following is provable: no protocol language or any kind of superalgorithm language can be written which will solve all problems in all circumstances, or even produce always identical results in different hardware. As for iterative procedures and algorithms in general, there will be cases where the result of the protocol is fairly insensitive to the details of the protocol at a given location in the protocol, and cases when the results are incredibly sensitive to the nature of the protocol at that point and even to the precision of the numbers as implied by the hardware supporting the computation. The proof is identical to those arguments demonstrating 'chaos' as sensitivity to starting conditions for any system or simulation of a system. Because of the importance of this, the effects of changing protocols in various ways should always be explored.

To facilitate this and for other obvious reasons, GLOBAL must also be an interactive language when interactive use is required. With necessary limitations both fundamental and practical on the limits of any current approach, GLOBAL is also designed to be open-ended and capable of growth: at the very least, GLOBAL should make such development and testing of new programs and protocols as easy as possible.

3 METHODS

It is meaningful to talk about the method of developing GLOBAL and GLOBAL itself in regard to methodology, since in fact GLOBAL was used as the tool to develop GLOBAL further. To clarify this and to show the potential of future growth for GLOBAL, we shall outline the historical steps of development.

Initially, a minimal system of capabilities was developed with the following properties.

(A) First steps

(1) A statement represents an interpretable and executable string S bound by a dollar sign ($) and a semi-colon (;). A statement is only interpreted and executed when completed and statements are interpreted and executed in the order given.

(2) An embedded statement within another statement is allowed, namely,

$ . . . $ . . . ; . . . ;

and the innermost statement is interpreted and executed first when the outermost statement is completed.

(3) A statement embedded in another statement may leave a *trace* when executed. That is, the innermost statement $ . . . ; is replaced by a string of characters T in the outermost statement,

$ . . . $ . . . ; . . . ; → $ . . . T . . . ;

(4) In the case of most commands created, the trace is a null string. However, the statement $USER; unlocks the keyboard and accepts an input string. This string becomes the trace.

(5) Subsequently, the *evaluatable statement*

$LET (variable) = (constant, variable or expression);

is created, with three possible types of result and variable holding the result. These are *real, integer, character*. The result of the evaluation if real or integer is then converted to type character, represented by the numeric and number format characters. That is, after any assignment such as $LET I=I+COS(X); or $I=200; The statement $I; becomes the string "200". This becomes the *trace*. Curly brackets became a special case of embedded string designed for use as variable parts of command names and of variable names, for example. In the case of the embedded $ . . ; then string constants are normally reported and printed in quotes, and numbers are written out in a format with leading blanks. These are removed, so that

$SAY ORT {HELLO};

prints out ORTHELLO, whereas

$SAY ORT $HELLO; ;

prints out ORT"HELLO".

(6) The crucial statement $PROCESS was then created. This defines new commands and has the format

$PROCESS (new command name) = {(old statement)};

(old statement) may be *atomic*, i.e. one single existing statement, or system of embedding statements, or *compound*, i.e. a *sequence* of statements. The dollar sign is actually the indicator of the atomic statement while the semicolon is the general separator. Therefore, square brackets and a semicolon [. . .] ; surround the compound sequence of statements:

$PROCESS (new command name) = {[(old statement) . . .
(old statement)] ;} ;

Self editing and recursion are now allowed according to the formalisms of the Lamda calculus (see, for example, [5]). For example, recursion would be performed by

$PROCESS (new command name) = {[(old statement) . . .
(old statement) $(new command name);] ;};

At this stage we have the potential to develop a full programming system since the LAMDA calculus has long been demonstrated equivalent to the 'Turing machine'. However, user input, and evaluation of arithmetic and character strings has been built in at an even earlier and lower level via $USER, and $LET. . . ; This imparts great computational efficiency to the language and, via the use of the property of leaving the trace, allows rapid further development of the language.

(B) The monitor

In the above we did not say how statements are presented to the system. The user may well assume that they are input from the keyboard. For the purest form of GLOBAL this is not so. In fact, $USER; is the only fundamental command which can communicate with the keyboard at all. However, the whole system for communication between user and central routines has been written in GLOBAL, using $USER; in such a way that a greater variety of input modes seem possible. The part of GLOBAL which is coded in GLOBAL *and* which is the most important is the MONITOR.

The monitor monitors the processes and events which occur in the GLOBAL system. Events such as keyboard input, the results of a 'back end' calculation or messages from a communication network cause interrupts which lead to the interpretation and the appropriate action for each event.

The processes and events to be monitored must be declared to the system, using the GLOBAL command

$MONITOR MY_PROCESS, MY_TEST_AND_ACTION_PROC;

where MY_PROCESS and MY_TEST_AND_ACTION_PROC are examples of a process name and a procedure name where the procedure specifies a *protocol* for test-and-action. Thus $MONITOR creates a monitor process to check for messages from the process name. When a message is received, the protocol implied by the procedure name is executed. This protocol is, of course, the GLOBAL code referred to above.

Not everything is written in GLOBAL, of course. For speed, the core routines for such operations as molecular mechanics have been written in FORTRAN 77. In cases of extreme 'fine tuning' to specific devices we employ not only special vectorization routines but in some cases we also have the most time-consuming steps written in micro-code. FORTRAN 77 was chosen for portability (since it still remains the world's best dispersed compiler) and called by GLOBAL. However, it greatly decreased the number of commands that had to be encoded in FORTRAN 77. The pertinent point for present purposes is that although the monitor is a 'master procedure' written in GLOBAL, it does in a later phase of development have some commands which evoke extensive code written in FORTRAN: these are primarily to do calculations involved in rapid communication between processes, which requires some explanation.

(C) Implementation of higher program structures

At this stage of truly communicating with GLOBAL via the keyboard the problem of designing a language both structured and interactive was tackled. BEGIN and END brackets were implemented with the same functions as in the ALGOL/PASCAL group of languages. However, they are formally separate statements in their own right: $BEGIN; and $END;.

The conditional action structures

> $IF (logical expression) ; (statement)
>
> $IF (logical expression) ; $BEGIN; (statement) . . .
> (statement); $END;

were then implemented, followed later with $ELSE; $ELSIF; capability.

The loop control facility has been structured in the same way:

> $DO (variable), (start value), (stop value), (increment);
>
> $BEGIN; (statement) . . . (statement) ; $END;

The logic is that the following statement is executed while the implied condition is true, or the following set of statements of the following statement is $BEGIN; The $ELSE; statement is formally equivalent to the $IF but takes the complement of the evaluated truth value in the $IF test, recorded at that hierarchical level. Thus $IF . . . $ELSE; and $IF . . . $ELSEIF . . . can be envisaged as brackets which can be nested as can IF, ENDIF in FORTRAN. However, there is a formal difference between

> $IF (logical expression); (statement 1) (statement 2)
> $ELSE; (statement 3)

and

> $IF (logical expression); $BEGIN; (statement 1) (statement 2)
> $END; $ELSE; (statement 3)

in that in the first case (statement 2) will always be executed. It is only the $BEGIN; $END; brackets which really determine the hierarchal level. It is most important to note that there is a fundamental formal difference between hierarchal nesting at this level, and embedding $. . . $ . . . ; . . . ;. In particular, the innermost $BEGIN; . . . $END; is *not* executed first. However, a series of the given level statement is not interpreted and executed until the closing $END; is accounted at AND if the level is not dependent on a condition which is false. This reasonable action should be remembered since although GLOBAL is a structured language, it is also interactive. The potential difficulties arise in that structure implies a hierarchal arrangement, while interaction with a human being is purely a linear sequence of events. Difficulties are avoided by this approach.

In most instances one can at least think simply in terms of programming for calculations which follow sequentially in time. Even machines such as the current Cray can be envisaged for conceptual purposes as being serial; it is simply that vectors can be handled with operations on all elements being more or less simultaneous (in practice, in most vector machines they are 'pipelined'). This is not conceptually difficult since in school we are taught to think of vectors as single entities on which we perform operations, e.g. vector A = vector B + vector C. However, in in-house versions being explored on highly parallel architecture such as the multiple transputer system, such thinking is not enough. It is necessary to program the choreography for the well-timed, integrated dance of whole processes, not necessarily in linear array. The $MONITOR command is also used to declare processes with particular relationships. The obvious one is in parallel:

$MONITOR ALL_TOGETHER_FOLKS, (PROC_A, JIMS_PROC, MARYS_PROC);

with choreography of communication managed either through specific channels or, very simply, by a master 'message board' electronic or file memory. In contrast, specific vector and pipeline operations such as vector A times vector B may have specific declaration types, simply stating they are arrays of data which can be treated together. A format which allows parallelism at the level of parallel $BEGIN;... $END; is being implemented at this time, and although it borrows unimaginatively from OCCAM it does have a few surprises, including useful commands such as $BEGIN WHEN process BEGINS;, $BEGIN WHEN process ENDS; $END WHEN process $ENDS, $BEGIN WHILE process RUNS; etc. Such extensions to $BEGIN; ... $END; detect basically four main states, (a) a specified process is running, (b) a specified process is not running, (c) a specified process has gone from not-running to running and a specific '$STARTED label;' command is not yet encountered in the specified process, (d) a specified process is hanging waiting for release by its own '$BEGIN WHEN' type of command. These commands are under active investigation and may be refined: the idea is not simply to produce flexibility and power, but simplicity of use.

Should the user require it, the most powerful notion is that processes are 'cellular automata' in a net. It is inappropriate to discuss this aspect here, but networks are fundamental to advanced GLOBAL Version 3 and beyond, and form the basis of programming for a whole variety of chemical applications (see sections G, H, and I). A chemically reacting system, transformations between different representations of chemical structure, conformation and function, and the linguistic considerations of user-friendly control of chemical calculation are all different manifestations of network thinking which are of direct application to a polymorphic programming environment for chemists.

(D) Development of expert system

The present version of GLOBAL has many ways of approaching the expert system capability. For one approach the essentials were already in position at the conclusion of the 'first steps' stage (A). Since evaluatable commands leave a *trace*, and the trace required is computed at the time of evaluation of the innermost embedded statement as in $... $...;...;, then any part of a GLOBAL statement can be variable. In particular the real, integer, and character variables can be *variable-by-name* as well as variable-by-value.

The proposition ALL_MEN_ARE_MORTAL in GLOBAL is simply a variable, and the truth value can be binary if defined as type integer, and probabilistic if defined as type real.

The predicate ALL_MEN_ARE_$LET SOMETHING=SOMETHING is a variable also, which is defined explicitly at interpretation and evaluation time by first evaluation of STRING and leaving the trace which is the characters contained in the variable called SOMETHING. In practice, of course, a special readable and computationally efficient format is developed, in which curly brackets ({...}) always force an evaluation (whether or not an assignment is made within the curly brackets). The predicate then reads:

ALL_MEN_ARE_{SOMETHING}

If this predicate is declared first, then subsequently the proposition

ALL_MEN_ARE_MORTAL

then the binding between all occurrence of SOMETHING to MORTAL which is required for a predicate calculus is achieved by an *implied* assignment

SOMETHING = MORTAL

made automatically and invisible at the time of entering the proposition.

Subsequently, relational and logical type operations are defined as mathematical functions and in ways which lend themselves to 'functional programming'. Let v be a variable, constant or expression, including a proposition or predicate, then

NOT(v)

AND(v,v, . . .)

OR(v,v, . . .)

are defined according to the traditional rules of fuzzy logic.

Relational operators such as 'greater than' are defined to represent the degree of adherance to the relation tested, e.g.

GT(v,v,v . . . .)

is 1.0 if the arguments are in descending order of value, 0 if in rising order of value, and intermediate when in intermediate order.

There are several approaches to the connection of rules via IF. . . THEN constructs, and as shown above, one is a classical 'logical switch' IF-test. A functional approach can be taken to the IF-test too, however.

A function IF is defined to evaluate the upgrade of belief in v_2 given proposition v_1

IF(v_1,v_2,c)

and reads IF v_1 THEN v_2. The third (optional) parameter c is the degree of confidence in that rule – if omitted, 1.0 is assumed. These and other functions of logic and predicate calculus, VERY, NOTVERY, SOME, ALL are defined to have required behaviour and dependence or independence on order and nesting of use.

A decision process thus involves computation of the value associated with a number of variables as the consequence of a series of operations using the above functions, and with variables which are either propositions, or predicates which are bound at their time of being used as variables.

(E) Human language capability

The ability to parse and respond appropriately to human language input was implemented as follows. The data structure required for the $PROCESS statement was considered to represent a dictionary. So far, we have assumed that all words must ultimately chain through to GLOBAL statements.

However, it is perfectly possible to assign data to the data structure so that chaining via process does not terminate in an executable GLOBAL statement. Rather, it terminates in a word or set of words. A further data item is created which is a SYNTAX type, e.g. NOUN, VERB. Any GLOBAL input which terminates in such a word can then, via a program written initially in GLOBAL, parse the input string so as to recognize, for example,

ADJECTIVE of the ACCUSATIVE PHRASE or

NOUN of the DATIVE PHRASE linked to the SUBJECT PHRASE

The system is general and both dictionary and grammatical rules can readily be set to handle other languages. There is provision for some 100 types of grammatical phrase in principle. In fact, highly inflected languages such as Finnish which originally made little or no use of prepositions only have about twenty types of phrase by grammar.

The *meaning* of a sentence is only understandable in terms of the response of a computational system to that sentence. The response is defined in turn by the *protocol* written in GLOBAL and stored (usually but not necessarily) on the *monitor*.

The verb FOLD, for example, initiates the latest protocol for dealing with the 'folding' (three-dimensional modelling) of proteins. Such a protocol will in general make use of the above expert system capability.

The philosophy of the above approach is that GLOBAL and human language can be mixed: if the system cannot chain through to recognize a GLOBAL statement or procedure, then it attempts to parse.

(F) Formal commands

Most chemical commands follow the format for *formal* commands, i.e.

$COMMAND arg,arg,arg,arg, . . . ;

where arg is a constant, variable or expression of type real, integer, or character. The 'curly brackets' are not obligatory for the arguments even in the case of expressions. Missing arguments must be indicated by retaining the presence of the commas (,), except in the case when there are no further arguments reading from left to right (i.e., the right-most commas can be dropped).

The type of argument in the argument list should generally match that required by the command, for precisely the same reason that the type of parameters in the call of a procedure should match those in the definition. However, the arguments really are represented by the *trace* they leave (see above), and this gives some possibility for laxity. Integer and real values are freely interchangeable. A string constant must be in quotes, or it will appear to the system as a variable to be evaluated. Any arguments that are missing are assumed to be zero when a real or integer number is required, and a string of (80) blanks if type character is required. Normally this will imply that a default value should be used.

Though arrays cannot be passed through the argument list except as elements of arrays, an argument may represent many arguments by an *implicit repetition string*. This is indicated by round brackets (. . .) and such an argument has the format (arg,arg,arg, . . .) where again each argument can be a constant, variable, or expression of type real, integer, or character. The statement is then re-executed for each successive argument. Since more than one argument can be an implicit repetition string, this is very general and very powerful. For example, the editing command

$EDIT_LINE (10,20,30),("A/B/","/C/D/")

will perform both substitutions /A/B/ and /C/D/ (replace A by B and C by D respectively) on lines 10, 20, and 30. The resulting action, it may be noted, gives all combinations and is formally the 'relational product' of the implicit repetition strings given. For this reason one can also make a user-defined function such as MY_ENERGY (I,I+1,I+2),(J+1,J+2), J+3); which evaluates the energy between I and J+1, I+1, I+1 and J+2, . . . etc. and accumulates the sum to give the total energy.

(G) Transformation operations

The true power of these commands is that they apply transformations on data which cannot directly be seen, and which would be too complex to pass through an argument list as above. The argument list is primarily only for control of the transformation of ultimate interest. Typical transformations are:

Sequence of polymer units	⟷	Formula
Dihedral angles of rotatable bonds	⟷	Cartesian coordinates
Cartesian coordinates	⟶	Energy
Cartesian coordinates	⟶	Dynamics
Dynamics history	⟶	Statistical mechanical properties

However, these files are in human-readable format, and can be interrogated or edited by queries from GLOBAL.

The notion of transformation process and 'cellular automata' is linked in Version 2 GLOBAL, and it is important to consider why. A 'cellular automata' is a single process that can be assembled into a network which has far more complex properties, e.g. a transputer program, a biological cell in general, or a neurone in particular (see [6]).

The *overall* process of predicting the conformation of a protein via its structure would require a chain of operations (at least 2), starting rightmost as follows:

$$\{S\} \leftarrow F.P.\ \{R\}$$

Here {R} is the amino acid sequence, {S} a set of coordinates, and P an initial prediction process to provide a starting point for a folding simulation F. P is normally initially a secondary structure prediction, typically emphasizing short range (near-neighbour) effects. In the information theory approaches of Robson [3] P was formally an operator based on information-theoretic analysis of finite statistical data, {R} associated with experimentally determined conformation {S}. Literally, a code-breaking approach was applied, to determine P to the level that {S} could be predicted given a new code input {R} for which the translation {S} was not known in advance. A modern approach is that of neural nets, implemented in the BIOENGINE by specifying the array of 'cellular automata' and their linkages, then training as known {R} → {S} relationships in much the same way. However, the neural networks used in Version 2 GLOBAL are probably efficient and probably unique in the fact that they work backwards.

$$\{R\} \leftarrow p^{-1}\ \{S\}$$

That is, by the *inverse operation* P a sequence can be proposed from conformation.

The importance of this is as follows. Normally, *prediction* is said to be just *part of design*. It is necessary not just to predict {S} from {R}, but to discover the {R} which will give rise to {S}. The above approach simply states, however, that the problem is to predict {R} from {S}, and the *design is the inverse of prediction*.

In principle, this can be applied to genuine design of pharmaceutical properties, but in practice no realistic data has lead to the required conclusion.

Although the most interesting data for these conversions is transferred between files, the instructions for arranging the neural net automata can be explicit, e.g.

$NETLINK.,, P1 → P2, arg, arg, arg, . . . ;

where arguments arg depend on the cellular automata type. GLOBAL treats chemical reactors as just a special type of neural net:

$REACTION, , ENZ + SUB → COMPLEX, KFORWARD, KBACK;

These reactions are assembled internally to form a network, such as a metabolic pathway and flow between elements of the network is specified by, for example,

```
$C(J) = C(J) + C(J) * SUM_OVER J, (K(J)*DT)
              + SUM_OVER I, (K(I)*C(I)*DT);
```

where C(J) is the concentration of matter implied at cellular automata J corresponding to a chemical species, DT is a small time interval, and the K are the rate constants for incoming and outgoing processes.

Again, recall that the reason for the similarity is that the underlying structure of a neural net process and any other net process is the same. It is simply that the complexities, connections, weights, and in particular processes per cell are differently defined. Initially this is by use of an underlying primitive command from which $NETLINK and $REACTION are defined. In consequence, it is possible to have nets with mixed cellular automata as processes, a chemical and neural net, for example.

(H) The 'semantic net'

The above reminds us that the most sophisticated transformation operators, including the human brain itself, have network structure. The extreme case in an artificial intelligence system is the semantic net as an internal system modelling the relationships of the external world – the 'interconnectedness of all things' as author Douglas Adams has written. The BioEngine version as built at Proteus is being used to explore world-images as networks of relationships between things in a much smaller 'universe of discourse', the realm of chemistry. Perhaps the simplest example of a neural net and its importance in reasoning is in regard to the basic syllogisms of Aristotle: having been told that 'Aristotle is a man' and that 'All men are mortal', we know that 'Aristotle is mortal', and we know that 'Aristotle is not mortal' is untrue. That is, written as GLOBAL variables, the declaration of ARISTOTLE_IS_A_MAN and then ALL_MEN_ARE_MORTAL would automatically generate a new variable ARISTOTLE_IS_MORTAL with an associated value computable from the associated values of the two old variables. This is important. If POLYKILLIN_IS_TOXIC and POLYTHRILLIN_DEGRADES_TO_POLYKILLIN it would be useful if the expert system let us know that POLYTHRILLIN is a bit risky.

The simple syllogism illustrates just how complicated, subtle, and still open to interpretation in context semantic nets can be. If the 'nouns' such as ARISTOTLE and MORTAL(THINGS) are nodes of a net and the relations represent verbs or prepositions, then the syllogism forges a new arc connecting two nodes not originally linked. Noting the existence of relations such as SOME... ARE..., SOME... ARE... NOT..., NO... ARE... and various reorderings of possible arguments A, B, C which we can insert, so as to give, for example, SOME A ARE B AND SOME B ARE C, or SOME B ARE A AND SOME C ARE B, gives potentially 256 syllogisms and thus the problem is not completely trivial: however, only a small subset are valid syllogisms, i.e. would result in forging a new arc (do not waste time trying to prove that simple probability theory will yield the required solutions – IBM carried out research to prove it could not). Interestingly, the nature of this subset is not immutable in human opinion, or rather, we should say, there is a need to precisely define what we mean by ALL and SOME. Counter to first appraisal of the problem, it is ALL and not SOME which is most mutable. The medieval and modern logicians, for example, agree over what SOME means: SOME A means one or more (at least one) and is the existential qualifier because an A actually exists. ALL in medieval times also implied existence, but in modern logic this is not so. ALL A ARE B means IF A EXISTS, THEN IT IS B (but it may not exist!). The interpretation of the input by the user is then of paramount importance, because if his intent is the same as that of the older logicians, then there are more arcs to forge.

There are several reasons why it is undesirable to propose an inflexible solution to semantic nets: there are still some unanswered questions about the best way to proceed. Are verbs just rather more active prepositions (which also show the relationships between phrases) or are they operators which transform nets? Should the connecting arcs not be quantitative rather than binary? If so, how may one efficiently compute the effect of a perturbation on one part of the net, i.e. of the effect of a changed degree of confidence about the relation between A and B on the relation between X and Y. Indeed, can this be a problem in any event? In practice, these semantic nets are constructed as protocols in GLOBAL using the IF functions to forge the semantic net. However, research into the area of nets and the perturbation of nets has rather more relevant applications to chemistry and molecular biology than simply to the expert system *per se*, as discussed in the final section.

(I) Molecular structure

Here we consider (somewhat but not entirely abitrarily) an example of treating a problem of medical and commercial importance. The example is chosen to illustrate how a variety of different aspects come together in treating a practical problem, or of even recognizing it exists, and also to illustrate how different philosophies of the BioEngine can come together – so justifying the use of a large unified PPE system. The specific example would answer the question 'What can networks have to do with conservation laws – and how can one build a computation protocol out of the relation?'

Can it be argued that research into the stability of networks is relevant to chemistry? In fact, of course, there are many important examples of network systems which are chemical, and many more systems which do not appear to be networks but which can be treated as such. Polymer networks are an obvious example, certain problems arising in protein engineering are less so.

Consider modelling the three-dimensional structure of a protein from the known three-dimensional structure of a closely related homologous protein. In such a case, a few units (residues) of the polymer chain will differ. These are involved with hydrogen bonding interactions with the sidechain and backbone of other units, and/or their own backbones. Suppose one of the differences means that a group A which normally donates a hydrogen bond to group B in the protein of known three-dimensional structure cannot do so in the protein being modelled, because the required hydrogen bond receptor group is missing. Since in protein structure there is a rule-of-thumb that all hydrogen bonds that could be made *are* made (though many will be to the solvent), the unsatisfied hydrogen bond of A will try to participate in hydrogen bonding elsewhere. It can do so with C, but only by displacing the original hydrogen bond donor D, which must then donate elsewhere. How is it possible to deduce the hydrogen bonding of the protein of unknown three-dimensional structure in such a case?

Now we come to the key question: does the chemical difference such as would arise in the mutation of one amino acid to another lead to locally accommodating the new hydrogen bonding, or does the effect ramify through the protein, causing extensive rearrangement of the network? If the latter can occur at least sometimes, when can its occurrence be predicted?

Networks can be prone to catastrophic changes in response to perturbation because there are so many cross-influences. Though equally they can be designed to resist change (enthalpically like a diamond, or entropically like a rubber), one cannot escape from the fact that even very ordered lattices in three-dimensional space can undergo the catastrophic global change known as the phase transition. Be that as it may, the situation is still relatively simple with two states and a single sharp transition. Some systems are much more complex. Amongst the hardest of the systems for the statistical mechanicians to enumerate states and derive properties for are the 'spin glasses'. In fact, of several possible approaches to the problem of protein hydrogen bonding the most obvious perhaps is the general 'spin glass' approach (see, for example, [7]), because both true spin glass systems and hydrogen bond networks are networks of dipoles. The systems generally studied as analogous to spin glasses also have many relationships with the protein problem. There are also some differences, and the matter of true analogy rather depends on the definition of the phenomenon of 'frustration'.

There is certainly a class of frustration which arises in the modelling context. Suppose we invoke a conservation law for an extrinsic function of conformation – the fraction of intramolecular hydrogen bonding, say 60%.

```
$CONSERVE FRACTION_H_BONDS=0.6;
```

The remaining 40%, approximately, will involve the solvent. We then run dynamics such that this fraction is conserved. By focussing the search only on regions which conserve the degree of hydrogen bonding this method is a very fast way to answer the question: are there equally good solutions elsewhere? By adapting the ingenious code of Rodney Cotterill for solving problems of this class rapidly, this becomes a particularly useful application of the conservation approach discussed in THEORY. Or is it? The effect, as discussed in section 2 is that the simulation has trajectories in phase space which are confined to a manifold with one dimension less than the phase space. Now it is possible to prove by probing the simulation that this manifold of 'constant hydrogen bonding fraction' is finite, unbounded, and *discontinuous*. There is a gap between solutions

FRACTION where _H_BONDS = 0.6 which correspond to a different value and which thus cannot be crossed without breaking the conservation law in force. That is, while the simulation explores solutions which give FRACTION_H_BONDS = 0.6 extremely rapidly, it is *locally confined* and in a general sense *frustrated*. This general sense is that it cannot reach an alternative, equally satisfactory solution without making a magic jump to a quite different region of phase space. A perturbation introduced by relaxing or changing the conservation law, say

$CONSERVE FRACTION_H_BONDS = 0.5;

can sometimes cause a quite catastrophic change in hydrogen bonding and indeed other conformational aspects by an escape of the trajectory to the different region of phase space.

It is easy to see that, while such problems in modelling are still difficult to overcome, an understanding of the nature of the problem is desirable. In the worst case, one might 'number crunch' forever and carry out an efficient and exhaustive search – which is local only. Moreover, a computational means of harnessing that understanding is essential. In fact there is a computational approach (indeed more than one) to resolving most efficiently and speedily the above problem, and we shall leave the reader the puzzle of deciding what that approach would be. A hint is that to do it properly and robustly requires a moderately complex protocol.

4 RESULTS

Above we noted that the method of developing GLOBAL, and GLOBAL itself, is a suitable topic for section 3. As it happens, GLOBAL is also the *result* of the development process. The question of the quality of result which we consider here can be measured at several levels

(i) Has a unified system involving many different aspects of theoretical chemistry been achieved as an integrated and coherent whole?
(ii) Is it easy to use, interactively and in batch?
(iii) Can it lead to protocols which solve problems?

We argue that the answer is yes to all three, though the third requires some comment. The principle answer is twofold. Firstly, it has been subjected to 'blind tests' by the pharmaceutical industry in which the answer was known, but not to the user. These will be announced in the near future. Secondly, it has generated useful patents in a few months of operation, particularly in the area of vaccines and diagnostics. Some five patents that we know of, and the basis of some five more, have already been generated in house or for clients via contract work. Most important of all, it has been a research project into the nature of the polymorphic programming environment which is demanded by chemists and which has revealed a variety of directions in which the future of controlling and using advanced calculational chemistry must lie.

We are grateful for the participation of those pharmaceutical companies who spent a great deal of their time to advise us as to the general form which such a system should take. Although we have emphasized the polymorphic environment rather than simulation algorithms as such, developmental testing of a supercomputer system for running simulations requires enormous amounts of computer time, certainly many hundreds of hours per month. We are therefore also grateful to Cray, Control Data Corporation, IBM, Convex, Alliant and Norsk Data who all variously provided support, advice, scientific input and extensive amounts of computer time. All machines performed excellently. We are also grateful to ETA Systems for providing our own ETA 10 mainframe supercomputer; the fact that this machine has now been withdrawn from the marketplace should not detract from the excellent character of the arrangement, which was developed in earlier days when ETA was in its prime, nor the excellent technical capabilities of what was a remarkable machine.

REFERENCES

[1] Robson, B. and Garnier, J. *Introduction to Proteins and Protein Engineering* Elsevier Press (1986, 1987).

[2] Dirac, G. *Proc. R. Acad. Sci. Series A,* **123** 714 (1929).

[3] Robson, B. *Biochem. J.* **141**, 853–867, (1974).

[4] Nelder, J.A. and Mead, R. *Comp. J.* 7, 308–313, (1965).

[5] Michaelson, G. *An Introduction to Functional Programming through Lambda Calculus*, Addison-Wesley (1989).

[6] Fogeman, A. *et al.* (Eds) *Automata Networks in Computer Science – Theory and Applications* Manchester University Press (1987).

[7] Stein, D. *Scientific American,* **261**, 36–42, (1989).

[8] Dutta, A. *The Explicit Support of Human Reasoning in Decision Support Systems* Academic Press (1987).

The growing use of topological indices for property prediction

D.H. Rouvray, Department of Chemistry, University of Georgia, Athens, Georgia 30602, United States of America.

Topological indices (TIs) are mathematical descriptors used to characterize numerically the molecular architecture of chemical species. In recent years TIs have come to the fore because of a growing awareness that they can provide a powerful means of predicting molecular behaviour. TIs are now known to correlate with an extensive range of molecular properties, including physical, thermodynamic, chemical, biochemical, biological and medicinal properties. In the past decade alone, the ever increasing repertoire of novel applications has embraced the design of pharmaceutical drugs, the prediction of carcinogenic behaviour, the characterization of polymeric systems, the optimization of air-breathing fuels, and the study of crystalline materials. Here we present a brief introduction to TIs and discuss a number of the current and potential applications of the indices in the chemical domain.

GENERAL INTRODUCTION

The chemical structures that we encounter in such rich abundance on our planet have been in existence for a comparatively short time in cosmological terms [1, 2]. Chemical structures can be formed only when individual atoms or groups of atoms interact under fairly stringent conditions. Modern cosmological theories [1, 2] indicate that such conditions first prevailed on Earth about five $\times 10^9$ years ago. Cosmic matter is thought to have been generated initially in the form of hydrogen and helium approximately 15 $\times 10^9$ years ago during the cataclysmic inferno usually referred to as the Big Bang. It took roughly another 10×10^9 years of roasting of these two primordial elements in the interiors of stars to produce the range of heavier elements found on Earth today. After a brief cooling-off period, the new elements began to react to form simple compounds such as water, ammonia and methane. These species in turn reacted together to yield ever more complex structures such as amino acids, proteins and nucleotides. The whole process culminated in the appearance of a wide variety of living organisms. The earliest life forms are believed to have evolved from a soup of such compounds around 3.8×10^9 years ago [3]. In Table 1 several of the major milestones in the development of chemical structures on Earth are listed.

Table 1 – The emergence of chemical structure on earth

Time from present (10^9 years)	Evolutionary development
15	The 'Big Bang' followed by the expansion of cosmic matter
14	Formation of heavier atoms (C, N and O) in star interiors from the primordial elements (H, He)
4.6	Formation of the Solar System and the planet Earth
3.9	Formation of simple molecules (H_2O, NH_3, CH_4, etc.) on Earth
3.9	Formation of complex molecules (amino acids, proteins, nucleotides, etc.) on Earth
3.8	Appearance of the first life forms on Earth
3.5	Formation of earliest known stromalites
2.8	Fossil evidence for photosynthesis in blue-green algae on Earth
1.6	Appearance of eukaryotic cells
0.7	Appearance of multicellular aquatic species

The precise conditions necessary to permit the formation of complex structures and living systems on our planet have been investigated in some detail [4]. Such studies have indicated that a large number of seeming coincidences had to occur. For instance, the physical constants of nature needed to be in certain, well-defined ratios to one another, e.g. the ratio of the strengths of the electric and gravitational forces had to equal the square root of the number of atoms in the observable universe. Moreover, the universe itself had to be of exactly the right size, as one smaller than ours could not have generated the elements necessary for the construction of complex structures. It has become increasingly apparent that the global structure of the universe is intimately interlinked with the conditions required for the emergence of complex structures and the sustenance of life forms on Earth. This hypothesis, first enunciated by Dicke [5] in 1957, is nowadays referred to as the Weak Anthropic Principle. The philosophical and scientific implications of this and its much more contentious twin, the Strong Anthropic Principle, – which maintains that the complex structures on Earth were bound to evolve in the way they did – have been examined in two recent monographs [6, 7].

The role played by structure in determining the behaviour of chemical substances first began to be explored in earnest during the middle part of the nineteenth century. It was at this time that the foundations were laid for what is now usually referred to as structure theory. From the outset structure theory concerned itself with the types and nature of the structures formed by organic molecules [8]. Thus, homologous series were studied and defined by general chemical formulas, the benzene ring was investigated and shown to possess a hexagonal structure, and the saturated carbon atom was described in terms of a tetrahedral structure. The term chemical structure itself was first introduced in 1861 by the Russian chemist Butlerov [9], and the structural formula was first employed in the same year by the Scottish scientist Crum Brown [10]. A number of the major accomplishments of structure theory in its heyday period of 1845–1875 are listed in Table 2. It should be emphasized that structure theory was to have a very profound impact on the future development of both chemistry and chemical industry. Because it gave a detailed insight into the microworld of chemical structure it was able to revolutionize our understanding of chemical processes and reactions. Indeed, one science historian has gone so far as to assert [11] that structure theory 'may be the most fruitful conceptual scheme in all of the history of science'.

Table 2 – Some milestones in the development of structure theory.

Principal researcher	Year	Contribution to structure theory
Charles Gerhardt	1845	The concept of homologous series
Auguste Laurent	1854	Diagram of the benzene ring
August Kekulé	1858	General formula for the alkane series
Archibald Couper	1858	Use of a line to represent a chemical bond
Alexander Crum Brown	1861	Use of the structural chemical formula
		Use of the double line to represent a double bond
Alexander Butlerov	1861	Use of the term chemical structure
Edward Fraukland	1866	Use of unattached valence bonds
August Kekulé	1867	Representation of the tetrahedral carbon atom
Emil Erlenmeyer	1867	Use of the triple line to represent the triple bond
Flavian Flavitsky	1871	Enumeration of the alcohol series
Arthur Cayley	1874	Enumeration of the alkane series
Jacobus van't Hoff	1874	Enumeration of stereoisomers

The notion that the structure of the molecular skeleton of a chemical species is of decisive importance in determining the bulk properties of the species dates from the 1850s. Until that time, chemical species had been represented by structural formulas that depicted only the so-called 'chemical' positions of the atoms in the molecule. The designation 'chemical' meant any arbitrary constellation of atoms that could plausibly account for the observed behaviour of the species. The 'chemical' positions of atoms were used because it was widely believed that the physical positions of the atoms in space were unknowable. During the 1860s, however, the situation was to change and formulas depicting the physical positions of the atoms began to appear [12]. The change was complete a few decades later and it then became accepted chemical orthodoxy that every stable molecule could be represented by an atomic architecture that was constructed in a well-defined and determinable fashion. The precise manner in which the atoms were interconnected in space came to be described as the connectivity or the topology of the species [12]. It is our prime thesis here that the fundamental importance of topology in determining the behaviour of chemical species has only begun to be fully appreciated in comparatively recent times.

THE ROLE OF TOPOLOGY

Three different avenues of approach have traditionally been explored in the search for methods of predicting the behaviour of chemical species from their molecular structure. These three approaches may be described as the (i) topological, (ii) geometric, and (iii) quantum-chemical approaches. The first and oldest of these approaches takes as its starting point the connectivity of molecular species and emphasizes the existence of linkages between atoms rather than the precise nature of the links. This approach has been extensively used for the prediction of the so-called additive properties of species, such as refractivity and heat of formation, which are roughly additive in series of related compounds such as the members of homologous series [12, 13]. The geometric approach concerns itself with the precise geometry of molecular species and highlights the bond lengths and bond angles. This approach is exemplified by studies on the chiral nature of the asymmetric carbon atom and on the conformational features of organic molecules [14, 15]. The quantum approach, which dates from the second decade of this century, focusses on the bonding holding species together and attempts to model the electronic interactions occurring between the atoms in a structure. This approach has given rise to the powerful discipline of quantum chemistry [16, 17] which is widely regarded as affording the best hope for the accurate prediction of molecular behaviour.

With the increasing prominence of quantum chemistry there has grown up a tendency to neglect topological approaches to both the modelling of chemical systems and the prediction of molecular properties. Because topological approaches rely on the non-metrical features of molecules, i.e. they ignore the geometrical features such as the bond lengths and angles, they have frequently been regarded as the poor relatives of quantum chemistry. Such attitudes are unfortunate, for the connectivity of a molecule actually embodies a wealth of information about the structure and nature of the molecule [18]. This information can be captured by topological descriptors, such as topological indices, which serve not only to characterize mathematically the structures of molecules but also provide us with a very valuable means of predicting bulk molecular behaviour. Endeavours spanning the past two decades to harness the information embraced by the connectivity have led to some remarkable accomplishments [19], about which more will be said later. For the present, it will be necessary first to discuss the relevant mathematical background. The primary mathematical tool we shall use for the description of chemical structure will be graph theory.

Graph theory is the discipline that studies mathematical objects known as graphs. The discipline is now over 250 years old [20] and has applications in virtually every science [21]. The graphs studied are not of the more common Cartesian variety but are rather devices used to characterize the connectivity of structures. A graph G may be formally defined as the pair:

$$G = (V, \hat{\Gamma}), \tag{1}$$

where V represents a set of points and $\hat{\Gamma}$ is an operator which maps members of set V onto themselves. In somewhat simpler language, a graph consists of a set of points (usually referred to as *vertices*), some or all of which are connected together by lines (usually referred to as *edges*). Graphs have been exploited to represent an exceedingly wide range of complex structures, ranging from family trees to urban transportation networks. In the chemical context, graphs have also been employed to depict many different types of structure, including that associated with molecular formulas, crystal lattices, and reaction networks [22, 23]. Examples of the graphs used to depict the molecules of methane and benzene are represented in Fig. 1.

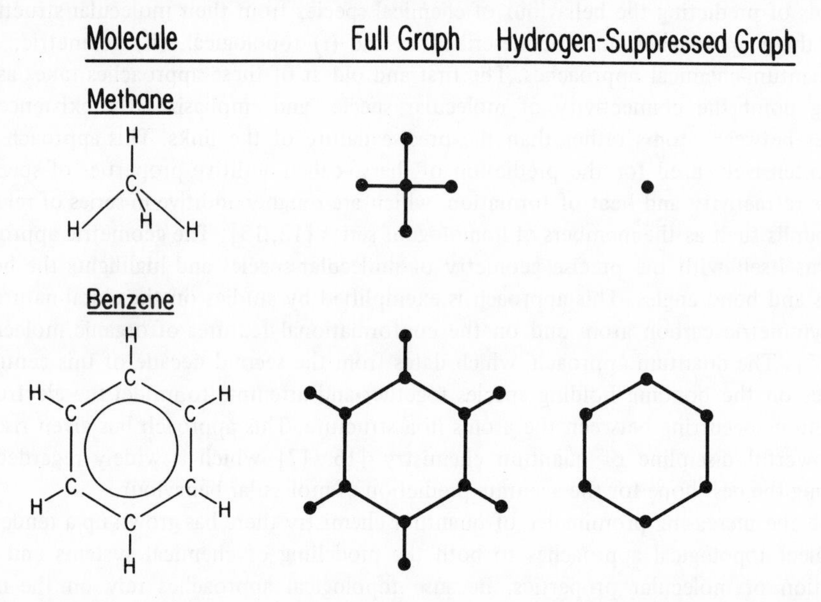

Fig. 1 – The full graph and the hydrogen-suppressed graph for the molecules of methane and benzene.

In order to keep the representation of molecular species as simple as possible, it is common to depict series by simplified chemical graphs. As is evident from Fig. 1, the hydrogen atoms in a molecule may be neglected and only the rump skeleton depicted. This can be done whenever the hydrogen atoms are not structure-determining, i.e. they are not involved in bridging but are located at the termini of the rump molecular skeleton. It should be mentioned as justification that hydrogen atoms are small in comparison to carbon or other first-row atoms and so do not contribute significantly to the electronic charge cloud that defines the size of first-row atoms. Another simplification that is frequently adopted is the depiction of multiple C–C and other bonds by a single graph edge. Such bonds vary greatly in their characteristics and may range from a single, electron-pair bond (ethane), through aromatic bonds (benzene), to triple bonds (acetylene). If the nature of the bonding needs to be emphasized, the graph edges can be weighted by appropriate factors, e.g. in the case of an aromatic bond by 3/2 or in the case of a double bond by 2. Weighting of graph vertices and edges is also frequently necessary if heteroatoms are present in the molecular structure. A variety of methods have been devised for carrying out the weighting in such cases, though none has yet gained universal acceptance. For further details on this the interested reader is referred to relevant texts [24–26].

TOPOLOGICAL MOLECULAR DESCRIPTORS

Once a molecular species has been represented by some appropriate graph (weighted or otherwise), the task in hand becomes one of converting this graph into the form of a serviceable structural descriptor. There are several different ways in which this may be accomplished. Methods based on the use among others of matrices, codes, sequences, Young diagrams, polynomials, and scalar graph invariants have been developed [27]. Our major focus here will be on the latter, as scalar descriptors afford an especially convenient means of characterizing graphs and thus the chemical species they depict. Starting from a given molecule M, the transformation that is made can be defined in terms of the following sequence:

$$M \longrightarrow G(M) \longrightarrow T(M) \qquad (2)$$

where $G(M)$ and $T(M)$ represent respectively the graph and the graph-based descriptor of the molecule M. This transformation effectively converts the rather intractable notion of chemical structure into a more tangible and useful descriptor. Such descriptors can be treated analogously to any experimentally determined property of the species [28].

Mathematical descriptors such as $T(M)$ are, however, not unique descriptors in general for the structures they characterize. That is to say, on occasion differing structures may be described in terms of the same descriptor. This is perhaps not surprising when it is remembered that at least in theory an astronomical number of different structures need to be characterized by descriptors. In the case of the alkane structural isomers, for instance, isomer counts of which are presented in Table 3, it is seen that the enumeration totals increase very rapidly. The $C_{40}H_{82}$ member has no less than 62 481 801 147 341 isomers, and this does not even include the stereoisomers! Compilations of isomer counts for a variety of other structural types reveal similarly enormous numbers for other series [29]. In spite of their lack of uniqueness, however, such descriptors provide extremely valuable tools for the chemist. In fact, this seeming disadvantage does not appear to impose any significant restriction on the practical application of these descriptors. When a choice of descriptors is made for a specific application, those which display as low a degeneracy as possible for the set of structures under consideration are usually selected.

Let us now discuss in somewhat greater detail the role of graph invariants, their mode of derivation, and their use in the characterization of molecular structure. Graph invariants are mathematical expressions that are characteristic of and derived from some graph. They are termed invariants because they are not changed by the manner in which the graph is labelled nor by its orientation in space. Simple examples of graph invariants are the number of vertices or the number of edges a graph contains. Whenever a graph invariant assumes the form of a scalar number, it is referred to in the chemical context as a topological index. In hydrogen-suppressed graphs of hydrocarbon species, the number

Table 3 – Structural isomer counts for members of the alkane homologous series, C_nH_{2n+2}.

Value of n	Isomer count
1	1
2	1
3	1
4	2
5	3
6	5
7	9
8	18
9	35
10	75
11	159
12	355
13	802
14	1858
15	4347
20	366 319
25	36 797 588
30	4 111 846 763
35	493 782 952 902
40	62 481 801 147 341
45	8 227 162 372 221 203
50	1 117 743 651 746 953 270

of vertices corresponds to the carbon number index and the number of edges to the number of carbon–carbon bonds in the species. Both of these quantities are thus topological indices, even though they are very simple conceptually and have been in existence for well over a century. The carbon number index, for instance, was introduced in 1844 by Kopp [30], but, because of its effectiveness, is still being used down to the present day [27, 31]. A major drawback with this particular index, however, is its very high degeneracy; all structural isomers with a given number of carbon atoms possess the same index. To overcome this difficulty, indices have been developed that can characterize branched species with a close approach to uniqueness [32]. An illustration of the procedure adopted in calculating the carbon number index for a hydrocarbon molecule is presented in Fig. 2.

As a rule, topological indices are determined by applying some mathematical algorithm to the chemical graph of the species under study. This means that topological indices can be calculated with complete accuracy, a situation that never obtains in the case of experimentally determined properties. After topological indices have been calculated for a test set of structures, regression analyses against experimental properties can be made and statistical correlations established. The precise conditions under which valid correlations can be achieved have been defined by Topliss and coworkers [33, 34]. The

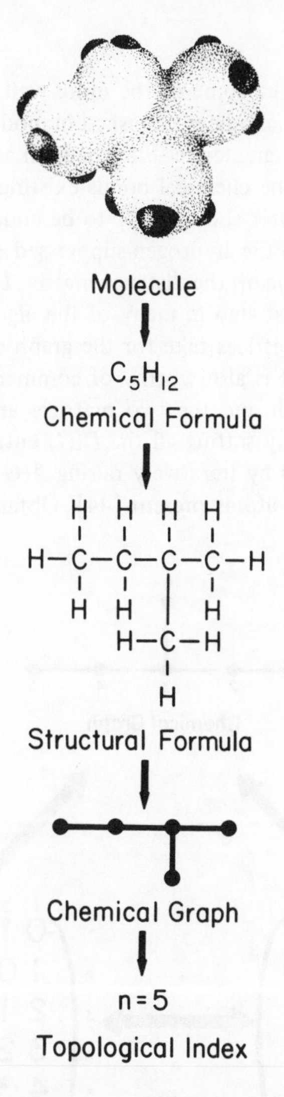

Fig. 2 – A schematic representation of the calculation of the carbon number topological index for the molecule of 2-methylbutane.

properties of structurally related species not in the test set can be estimated by calculating a topological index for them and then reading off from the relevant correlation curve the property of interest. This procedure has been employed for a vast array of different properties, including the physical, thermodynamic, chemical, biochemical, biological and medicinal properties of molecular species. In fact, topological indices have been so extensively used in recent years that they have become ubiquitous in their role of facilitating property prediction. At present some 120 topological indices have been put forward in the literature, though only a handful of these have been studied in any depth [27]. A number of reviews of topological indices and the correlations obtained with them have appeared in books [24, 25, 35], book chapters [23, 36, 37], and journals [19, 27, 38–41].

USES OF THE WIENER INDEX

We now turn to a consideration of some of the more well known and frequently used topological indices. The first of the indices published in modern times was that of Wiener [42] which was designed to characterize branched alkane species. This index was originally defined as the sum of the chemical bonds existing between all pairs of carbon atoms in the molecule, but was later shown [43] to be equal to one half the sum of the entries in the distance matrix of the hydrogen-suppressed graph of the molecule. It is interesting to note here that setting up the distance matrix, $D(G)$, or its close relative, the adjacency matrix, $A(G)$, is the first step in many of the algorithms used to derive topological indices. The form these matrices take for the graph of the molecule of 3-methyl pentane is illustrated in Fig. 3. It is also worthy of comment that the sets of arrows in this figure indicate that the graph and the two matrices are mutually interconvertible. $A(G)$ can be derived from $D(G)$ by setting all the $D(G)$ entries greater than one to zero; $D(G)$ can be obtained from $A(G)$ by iteratively raising $A(G)$ to the powers $2, 3, \ldots, n$, where n is the number of carbon atoms present [44]. Obtaining the graph G from either $A(G)$ or $D(G)$ is straightforward.

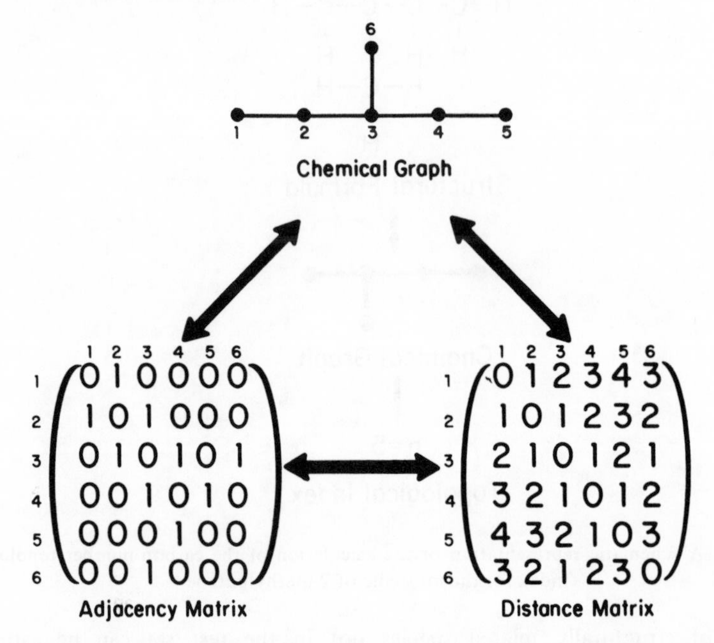

Fig. 3 – A depiction of the adjacency matrix and distance matrix for the molecule of 3-methylpentane showing the interconvertibility of the various representations.

From the above, we see that the Wiener index, $W(G)$, of some chemical graph G can be defined by the equation:

$$W(G) = \frac{1}{2} \sum_{i} \sum_{j} d_{ij}, \tag{3}$$

where d_{ij} symbolizes the ijth entry in the distance matrix, $D(G)$. This index has been studied in its own right because of its many interesting mathematical properties. Thus,

Plesnik [45] carried out an investigation on the mathematical characteristics of $W(G)$ for various classes of graphs. Wiener himself gave a general expression for $W(G)$ for graphs representing unbranched chains, i.e. for path graphs containing n atoms, he showed [42] that $W(G)$ always assumed the form $\frac{1}{6}(n^3 - n)$. This result and a number of other general, closed formulas for $W(G)$ are exhibited in Fig. 4. A very general expression for $W(G)$ from which it is possible to derive a closed formula for any tree graph, no matter how complex, was recently published by Canfield *et al.* [47]. The index can be extended to cyclic graphs following the proposal of Hosoya [43], who defined the d_{ij} entry in such cases to be the minimum distance between the vertices i and j. For a more detailed discussion on the mathematical properties of $W(G)$, the reader is referred to reviews by the present author [27, 44].

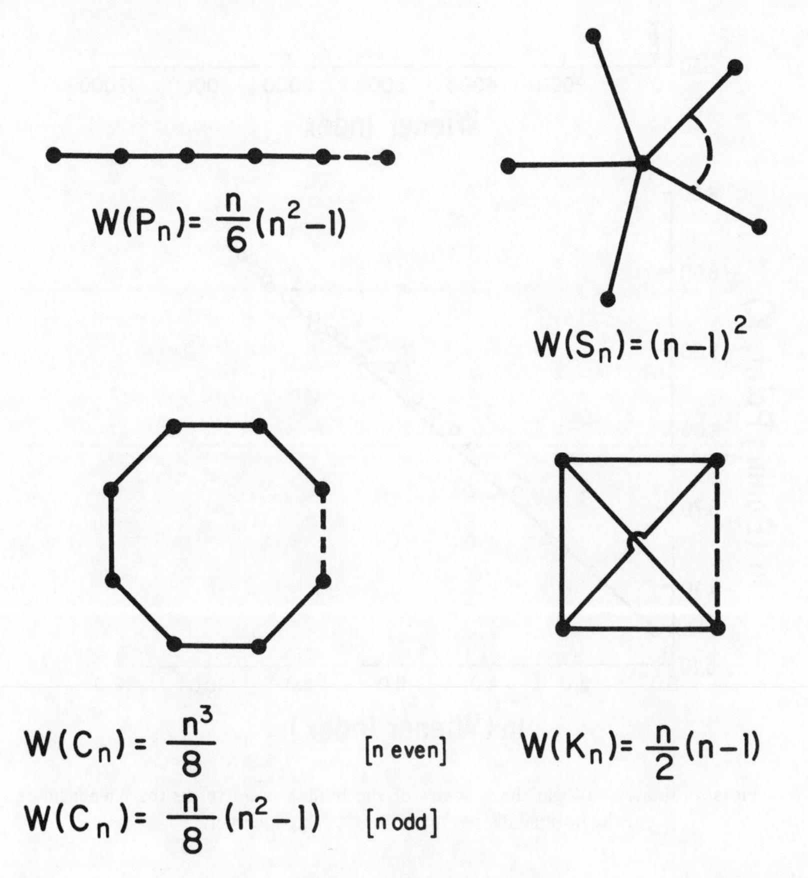

Fig. 4 – Illustration of some of the closed algebraic formulas obtained for the Wiener index for different classes of graphs.

If the Wiener index is plotted against some measured physicochemical property for the members of a homologous series, a curve similar to that shown in Fig. 5 results. In this instance, $W(G)$ is plotted against the boiling point of normal, i.e. straight chain, members of the alkane series. It is of interest that the pronounced curvature in this plot is considerably, though not completely, straightened when use is made of logarithmic scales. A variety of methods have been explcred to obtain completely linear plots. These

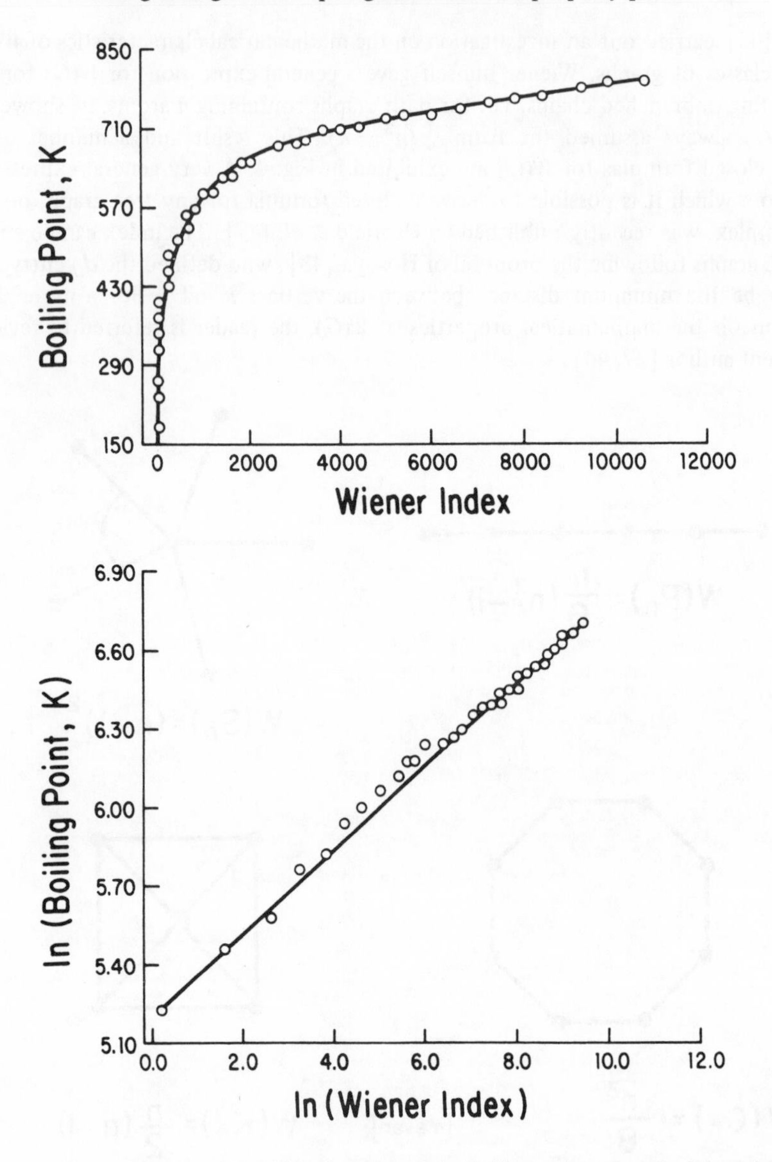

Fig. 5 – Plots on linear and logarithmic scales of the boiling point versus the Wiener index for the normal alkanes up to the $C_{40}H_{82}$ member.

have included the biparametric approach of Wiener [48], who used $W(G)$ together with another topological index in a linear combination, and the exponential approach of Walker [49], who introduced equations of the general type:

$$P = {}_{\alpha}[I]^{\beta}, \tag{4}$$

where P is some measured molecular property, I is a molecular descriptor, and α and β are constants. The two constants α and β can be obtained from the slope and intercept of logarithmic plots of P against I [50]. It is now realized that such plots cannot be straight lines because of the nature of the species involved. Thus Trinajstić [51] showed that the mean square radius, $\bar{R}^2(G)$, of normal alkane molecules is related to their Wiener index by the equation:

$$\bar{R}^2(G) = W(G)/n^2. \tag{5}$$

Furthermore, it was later demonstrated [52] that the curvature arises from the changing fractal dimensionality of the species. In fact, the curvature can be used to derive valuable information on the mean conformation adopted by alkane species at their boiling point [52].

In addition to modelling numerous physicochemical properties of both normal and branched members of homologous series, such as their boiling point [48], viscosity [50] or chromatographic retention time [53], $W(G)$ has found limited use in the biological realm where it has been used to model antibacterial activity [54] and narcotic behaviour [55]. Notable successes have also been achieved in predicting the properties of polymeric and solid-state materials. For the polymers, Mekenyan *et al.* [56] showed how it was possible to normalize $W(G)$ to yield finite values for infinite chains of monomeric units. The normalized $W(G)$ values were then employed to obtain good estimates of many polymer properties, such as melting point and refractive index, for systems ranging from polyethylene to polyethylene terephthalate [56]. For the solids, $W(G)$ has been used to study crystal vacancies [57]. A solid-state system can be regarded as being in its lowest energy state when $W(G)$ has its minimum value for the system. This approach has been applied to the study of vacancy migrations along preferred diffusion paths [57], the optimized positioning of double and triple vacancies in lattices [58], the optimal location of defect atoms in lattices [58], and the modelling of crystal growth processes [60]. $W(G)$ also exhibits correlations with the σ- and π-electronic energies in a variety of molecules from the arenes to spiro structures [61, 62]. On the basis of these correlations, it is possible to assess how structural changes in such systems will impact the electron affinity, first ionization potential, maximal light absorption, and electrical conductivity [63].

THE MOLECULAR CONNECTIVITY INDICES

The most popular of all the topological indices in terms of their frequency of application to date are the so-called molecular connectivity indices. These were elaborated from a single index put forward in 1975 by Randić [64]. In its original form the index was defined by the equation:

$$\chi(G) = \sum_{G \text{ edges}} (p_i p_j)^{-1/2}, \tag{6}$$

where $\chi(G)$ represents the index for the chemical graph G, p_i and p_j are the respective degrees of the adjacent pair of vertices i and j in G, and the summation extends over all the edges in G. The subsequent generalization of this index by Kier *et al.* [65] led to the development of the set of molecular connectivity indices as defined by the general equation:

$$^{h}\chi_r(G) = \sum_{k=1}^{\sigma_h} \prod_{i=1}^{h+1} (p_i)_k^{-1/2}, \tag{7}$$

where ${}^h\chi_r(G)$ is the molecular connectivity index of order h constructed for graph G using subgraphs of type r, h equals the number of edges in the subgraph type used, σ_h is the number of subgraphs of type r having h edges in G, and the subscript k extends over all the σ_h subgraphs. Equation (7) applies only to graphs in the form of trees, but may be readily modified for other classes of graphs [24].

As indicated above, the generalized molecular connectivity indices are defined in terms of certain types of subgraphs that exist within the graph, G, of the molecular species. To form these generalized indices, it was proposed [65] that four different types of subgraph be utilized, namely paths, clusters, path/clusters and cycles. Subgraphs other than paths are relevant only in the calculation of the higher-order molecular connectivity indices, i.e. those for which $h \geqslant 3$. Examples of the various kinds of subgraph are illustrated in Fig. 6. Because molecular connectivity indices are very often used to characterize molecules containing heteroatoms, special weighting procedures for both the vertices and edges involved were developed by Kier and Hall [24, 66]. Their basic idea was to correct for the valence of each heteroatom by subtracting from its number of valence electrons the number of attached hydrogen atoms [25, 26]. The corrected index has been universally denoted by the symbol ${}^h\chi_r^{\mathrm{v}}(G)$. Since this early work, a variety of other techniques for weighting graph vertices and edges have been put forward [24, 26, 27]. Although connectivity indices or orders ranging from 0 to 6 are in common use, the most widely used index is still that of order 1 ($h = 1$), which is, of course, precisely that originally introduced by Randić [64].

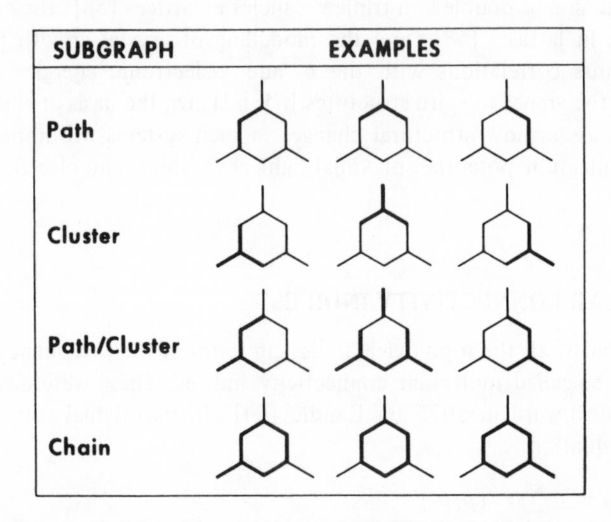

Fig. 6 – Illustration of the subgraphs used in deriving the higher order ($h \geqslant 3$) molecular connectivity indices. Note that the cycles are sometimes referred to as chains.

To summarize the manifold applications of molecular connectivity indices is an exceedingly difficult task because of their ubiquitous employment in most of the major areas of chemistry. Already, two entire monographs [24, 25] have been devoted to a discussion of these indices and their uses; in addition a fair number of review articles [19, 27, 38–41, 68] have also been written on the subject. We can therefore do no more here than present a very brief outline. Table 4 gives a number of typical examples of the

Table 4 – Correlations of properties with molecular connectivity indices

Bulk property	Correlation equation	Data points	Chemical species	Correlation coefficient	Reference
Physicochemical properties					
Boiling point	$t_B = 57.85\ {}^1\chi - 97.90$	51	Normal alkanes	0.9851	24
Partition coefficient (octanol/water)	$\log P = 1.48 + 0.950\ {}^1\chi^v$	138	Various organics	0.9860	24
Water solubility	$\log S = 6.702 - 2.666\ {}^1\chi$	51	Aliphatic alcohols	0.9870	24
Chromatographic retention time	$T_R = 482.12\ {}^1\chi - 559.60$	51	Various arenes	0.9911	69
Biochemical and biological properties					
Anaesthetic activity	$\log \frac{1}{c} = 4.376 - 3.729 \left(\frac{1}{{}^1\chi}\right)$	28	Aliphatic ethers	0.9790	70
Enzyme inhibition	$pC = 0.916\ {}^1\chi - 1.582$	13	Various organics	0.9660	71
Toxicity to fish	$\log \left(\frac{1}{EC_{50}}\right) = 0.749\ {}^1\chi^v - 5.630$	12	Organotins	0.9750	72
Soil sorption	$\log K_{om} = 0.550\ {}^1\chi + 0.450$	37	Various organics	0.9730	73

use of the $^1\chi$ index. The first four applications listed therein refer to physicochemical properties whereas the remaining four relate to biochemical or biological applications. The results reveal that excellent correlations are obtained for the physicochemical properties, with correlation coefficients exceeding 0.98, whereas for the biochemical and biological properties the correlation coefficients are somewhat lower, even though the data sets are generally much smaller. This trend is apparent in all the applications studied to date and has led to the use of correlations based on up to three descriptor variables to obtain improved results. In a multivariate study on 23 substituted amphetamines, whose hallucinogenic effect relative to mescaline was the property of interest, it was shown [74] for instance that the trivariate equation:

$$\log H = 45.16 \left(\frac{1}{^3\chi_p}\right) + 1.288\,^6\chi_p - 4.298 \left(\frac{1}{^4\chi^v_{pc}}\right) - 5.592 \tag{8}$$

yielded the best relationship, with a correlation coefficient of 0.920. There is now a growing tendency to combine molecular correctivity indices with other topological indices in multivariate correlations [37, 75].

At this point it is pertinent to enquire why topological indices correlate so well for such a multitude of quite different molecular properties. The short answer seems to be that the indices encode fundamentally important information about the structures they are used to characterize. In the case of the Wiener index, 90% of its value is known [76] to reflect the species volume and the remaining 10% its shape. The same holds true for the original Randić molecular connectivity index, $^1\chi(G)$ [76]. Edward [77] has suggested that the latter index correlates well with the properties of alkanes because it reflects both the numbers and types of different carbon atoms present as well as the mole fraction of the *gauche* conformations present in the species. Kier and Hall [25] have demonstrated that whereas $^0\chi(G)$ and $^2\chi(G)$ increase with the extent of branching in tree structures, $^1\chi(G)$ actually decreases. Such studies indicate that topological indices are able to reflect both the size and shape of molecules to varying degrees, with shape being of overriding significance for $W(G)$ and $^1\chi(G)$. Cramer [78, 79] established that it is precisely these two quantities which are of prime importance in determining the observed physical properties of many liquids. In fact he concluded that any molecular property that depends primarily on non-specific and non-covalent intermolecular interactions can be predicted from the topological structure of the species involved. Biological activity can thus be modelled with topological indices if the interactions are of a non-specific type [23], which implies that the behaviour of the substrate molecule at the biological receptor is essentially additive. Fortunately, a wide range of biological interactions appear to satisfy this criterion, thereby making the use of topological descriptors feasible in the biological context [26].

RECENT APPLICATIONS OF TOPOLOGICAL INDICES

In this section we focus on some of the more recent applications of topological indices. Although our emphasis here will still be on uses of the Wiener index and the molecular connectivity indices, we shall also make mention of certain other indices. We shall endeavour to highlight those applications which appear to hold great promise for the future utility of topological indices. One exciting development has been the use of data reduction techniques. These involve essentially the adoption of a simplified or reduced graph, G', rather than the full graph, G, for calculation of the indices. Such techniques are effective only if the reduced graph, G', retains the key structural information contained within the full graph, G. The conditions under which reduced graphs are equivalent in this sense to more complex graphs have been explored by El-Basil [80]. It has been shown for some time that polyhex graphs may be reduced to tree structures known as caterpillar graphs [81]. This fact was exploited in a study on polycyclic aromatic hydrocarbons in which $^1\chi$ indices were calculated for the original polyhex graphs and the reduced caterpillar trees. Both sets of indices were then separately correlated [82] against the positions of electronic absorption bands in the species. Good correlations were obtained for families of related structures [82] and it was found, rather surprisingly, that the correlations based on the reduced graphs were generally better than those based on the original graphs. Plots for the dibenzoacene and tribenzoacene families using reduced graphs are shown in Fig. 7.

Recent years have witnessed a growing use of clustering techniques for the classification of chemical structures [83]. A number of these techniques introduce similarity measures that are applied in the comparison of the structures. Because there is no absolute measure of similarity of chemical structures, however, a great diversity of different measures have been proposed [84]. Similarity-based techniques partition sets of molecules into mutually exclusive subsets by searching for similarities in their structural descriptors. The techniques all involve the three basic steps of (i) selecting appropriate variables to characterize the structures, (ii) weighting these variables, and (iii) defining some similarity measure. Although a wide variety of variables have been tried out to date, topological indices or combinations of these indices have found increasing favour in recent years [85]. The two major advantages of similarity-based techniques are that one is not restricted to the use of congeneric series of molecules or conformationally rigid species, and that the techniques afford a very valuable new route for the design of molecules that are optimized for specific applications. Such techniques have been adopted in the design of pharmaceutical drugs [86] and air-breathing hydrocarbon fuel molecules [87]. Fig. 8 shows the results of two similarity searches on a large database of bioactive molecules for the structures most similar to a specified structure. The variables used were structural fragments and topological indices. It is interesting to note that quite different outcomes are arrived at for differing variables.

Hydrocarbon fuels were also featured in a pioneering study [88] which demonstrated how topological indices could be applied to predict the sooting tendency of fuel molecules. A knowledge of this particular property is important for a variety of reasons, not least of which is that it provides us with a measure of the pollution potential of the fuel. The possibility of predicting the sooting tendency of a fuel molecule from its connectivity appears unpromising because the formation of soot involves a complex, multistage process in which the original fuel molecule is ultimately transformed into large carbon clusters, such as buckminsterfullerene [89]. However, it was proven [88] that the soot threshold index can be correlated with topological indices for a wide variety of different fuel molecules. The indices used were the Balaban averaged distance sum connectivity index [90] and the hydrogen deficiency index [91]. The former is defined for a graph G by the equation:

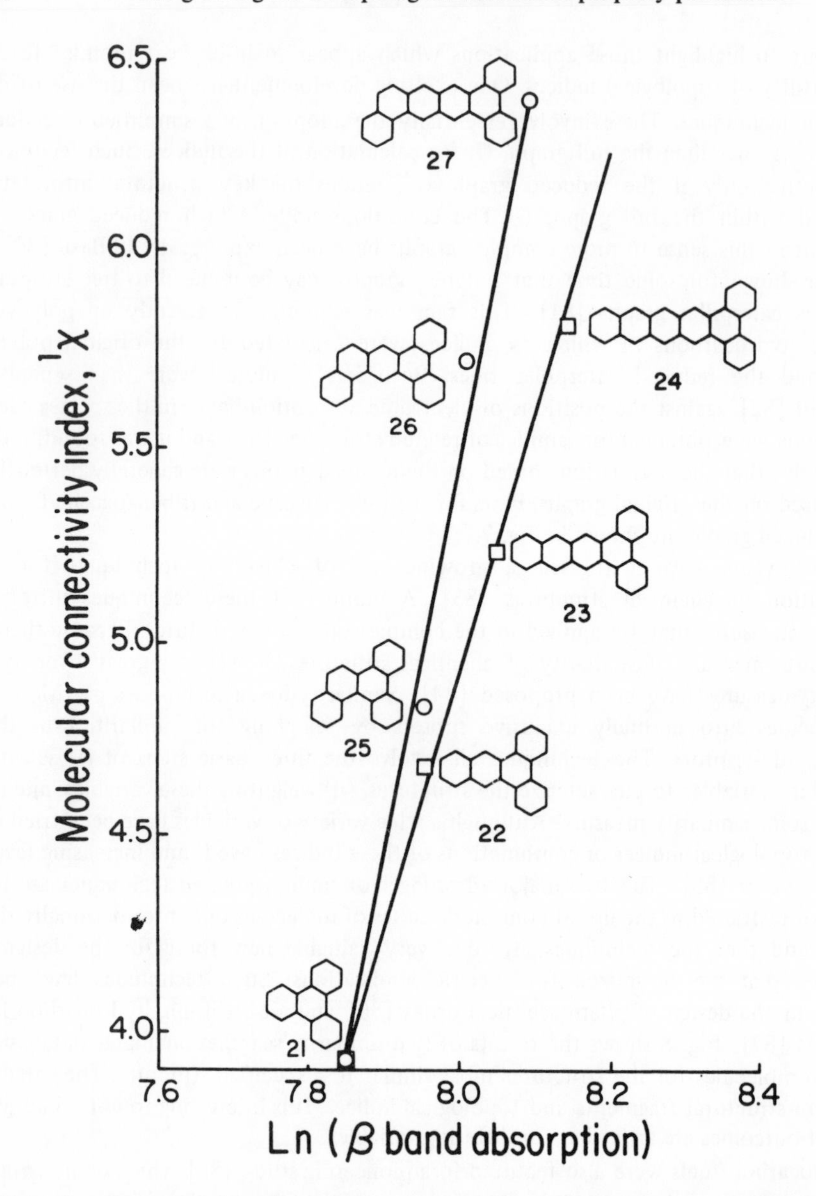

Fig. 7 – Plots of the $^1\chi$ molecular connectivity index against the logarithm of the β-band electronic absorption using caterpillar tree reduced graphs for members of the dibenzocenes (right) and tribenzoacenes (left).

$$J(G) = \frac{n_e}{\mu(G) + 1} \sum_{G \text{ edges}} (D_i D_j)^{-1/2}, \tag{9}$$

where n_e is the number of edges in G, $\mu(G)$ is the cyclomatic number of G, and D_i is the distance sum for vertex i in $D(G)$. The cyclomatic number, $\mu(G)$, is given by the expression:

$$\mu(G) = n_e - n_v + 1, \tag{10}$$

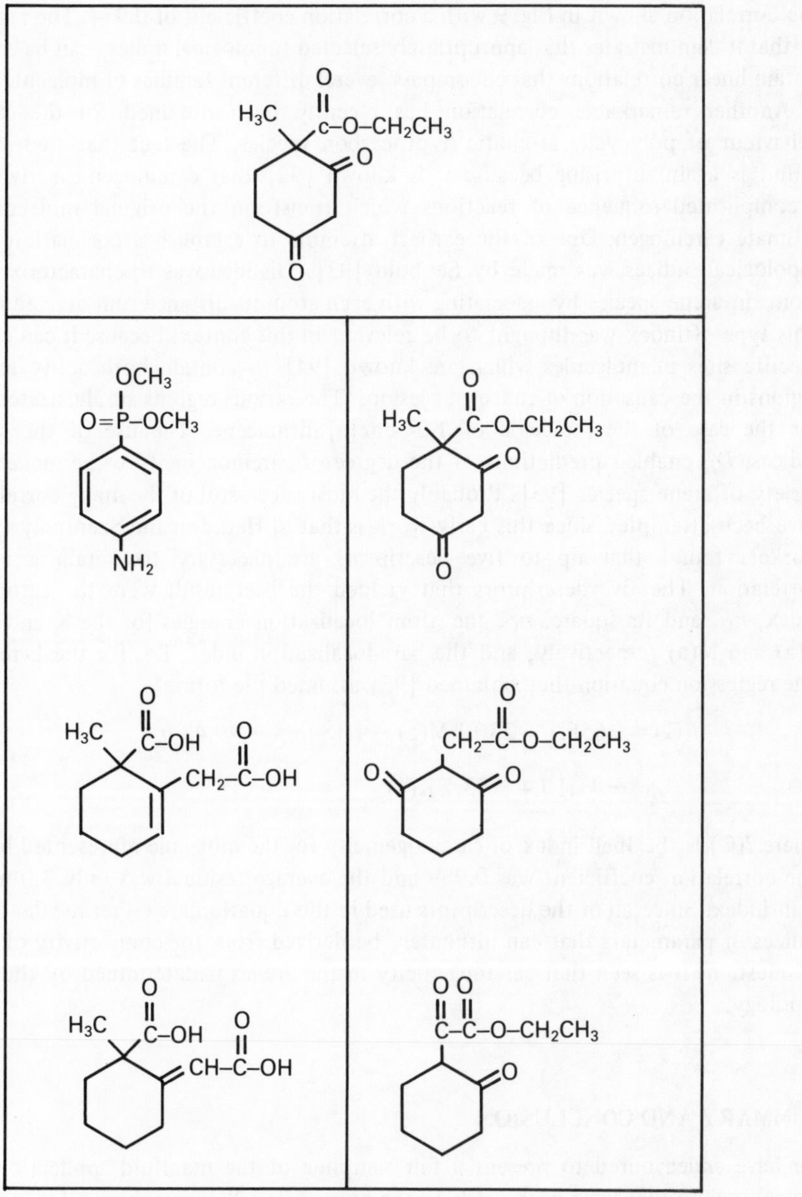

Fig. 8 – Similarity-based searches on a large database of structures to determine the most similar structure to that at the top. Note that the fragment and the topological index searches yield quite different results.

where n_v is the number of vertices in G. The cyclomatic number represents the number of independent cycles in G and corresponds to the chemists' hydrogen deficiency index which is defined as the sum of the number of rings and double bonds plus twice the number of triple bonds in a structure [91]. Plotting the product of $J(G)$ and $\mu(G)$ against the threshold soot index for 93 fuel molecules of varying degrees of unsaturation yielded the correlation shown in Fig. 9 with a correlation coefficient of 0.974. The result is novel in that it demonstrates that appropriately selected topological indices can be employed to obtain linear correlations that encompass several different families of molecules.

Another remarkable correlation has recently been obtained for the carcinogenic behaviour of polycyclic aromatic hydrocarbon species. The fact that a correlation was found is again surprising because it is known [92] that carcinogenic activity involves a complicated sequence of reactions which transform the original molecule into the ultimate carcinogen. One of the earliest attempts to establish a correlation using only topological indices was made by Seybold [93]. His idea was to characterize individual atoms in arene species by associating with each atom its distance sum derived from $D(G)$. This type of index was thought to be relevant in this context because it can characterize specific sites in molecules which are known [94] to contain both active and inactive regions in the causation of cancerous lesions. The various regions are illustrated in Fig. 10 for the case of the molecule of 1,2-benz[a]anthracene. The use of the site-specific indices, D_i, enabled predictions of the degree of carcinogenicity to be made for a wide variety of arene species [93]. Probably the most successful of the many correlations that have been attempted since this early work is that of Herndon and Szentpály [95]. These workers found that up to five descriptors are necessary to obtain a really good correlation. The five descriptors that yielded the best result were the carbon number index, n_C, and its square, n_C^2, the atom localization energies for the K and M regions, K(a) and M(a) respectively, and the para-localization index, L4, for the L region [95]. The regression equation they obtained [95] assumed the form:

$$I(G) = -655.1 + 230.1\,M(a) - 1.184\,n_C^2 + 49.43\,n_C - 41.1\,L4 - 73.7\,K(a), \qquad (11)$$

where I(G) is the Iball index of carcinogenicity for the molecule represented by graph G. The correlation coefficient was 0.989 and the average residual was only 3.0 units in the Iball index. Since all of the descriptors used in this equation are either regular topological indices or parameters that can ultimately be derived from the connectivity of the graphs in question, it is seen that carcinogenicity in the arenes is determined by the molecular topology.

SUMMARY AND CONCLUSION

We have endeavoured to present a fair sampling of the manifold applications of topological indices that have been made to the present day. We have shown, for instance, that topological indices are especially effective in correlating and predicting the physicochemical properties of molecular species. Typically, in this area correlation coefficients of greater than 0.98 are achieved. When we consider the biochemical or biological properties of species, however, the correlations are somewhat less impressive, with correlation coefficients commonly lying in the range 0.92–0.98. Even so, it is surprising that any correlations are obtained at all, for the properties in question usually involve a highly complex set of interactions at the molecular level. In such situations it might be anticipated that the interactions would be difficult to model with devices as comparatively simple as topological descriptors. The key factor here seems to be that

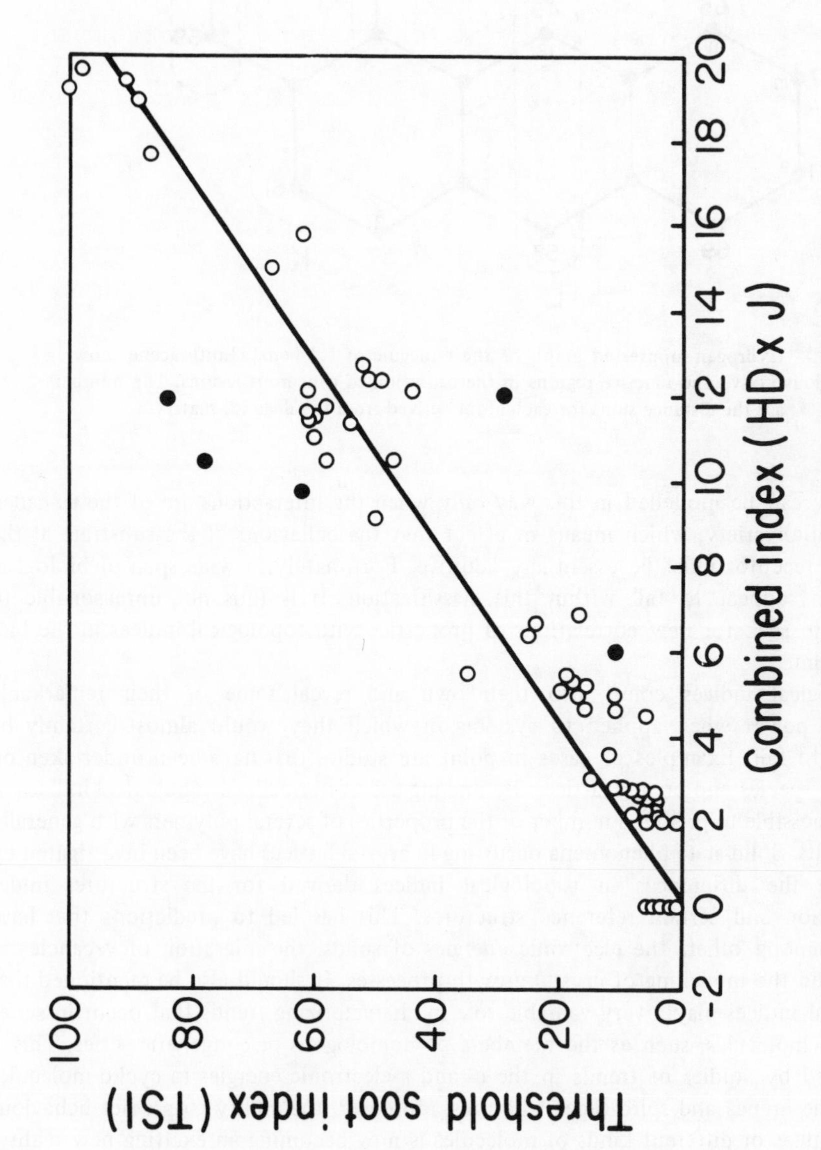

Fig. 9 – Plot of the threshold soot index against the composite index HDxJ for 93 hydrocarbon fuel molecules of varying types and structures. Black circles indicate the outliers.

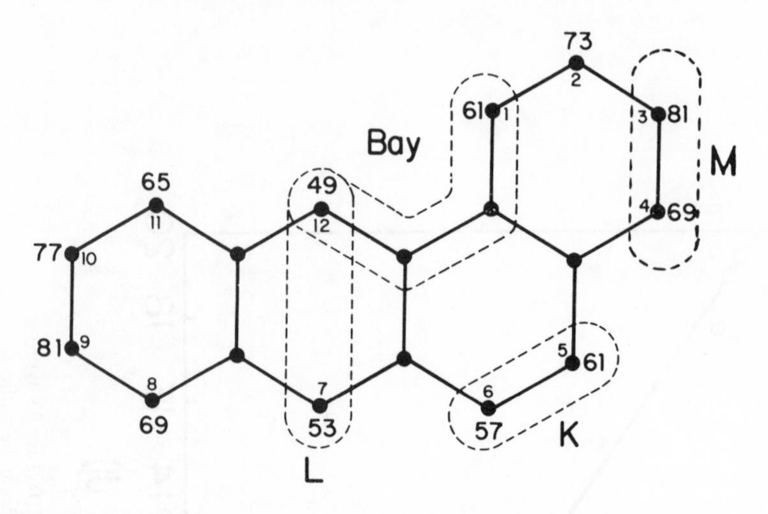

Fig. 10 – Hydrogen-suppressed graph of the molecule of 1,2 benz[a]anthracene showing the various active and inactive regions in the causation of cancerous lesions. The numbers are the distance sums for each atom derived from the distance matrix.

bioactivity can be modelled in this way only when the interactions are of the so-called 'non-specific' variety, which means in effect that the behaviour of the substrate at the biological receptor must be essentially additive. Fortunately, a wide span of biological interactions appear to fall within this classification. It is thus not unreasonable to continue to seek for new correlations of properties with topological indices in the biological realm.

Topological indices come into their own and reveal some of their remarkable predictive power when applied to systems in which they would almost certainly be expected to fail. Examples of cases in point are studies that have been undertaken on polymeric species and crystal lattices. By making use of normalized topological indices, it has been possible to predict a number of the properties of several polymers with generally good results. Solid-state phenomena occurring in crystal lattices have been investigated by examining the differences in topological indices derived for the structures under consideration and certain reference structures. This has led to predictions that have involved among others the electronic energies of solids, the migration of vacancies in lattices, and the modelling of crystal growth processes. It should also be mentioned that topological indices play a very valuable role in characterizing trends that occur in series of related molecules, such as the members of homologous or congeneric series. This is exemplified by studies of trends in the σ- and π-electronic energies in cyclic molecules such as the arenes and spiro-type structures. Moreover, the ability to predict behaviour across families of different kinds of molecules is now becoming an exciting new reality. Notable examples that have been published recently include the sooting tendency of many different types of hydrocarbon fuel molecules and the carcinogenicity of a variety of arene species. Developments of this kind are especially gratifying as cross-family correlations have traditionally been fraught with considerable difficulties. In summary, it may be justifiably said that topological indices currently provide us with very powerful tools for the prediction of a vast array of different molecular properties. Although we cannot know with certainty what the future holds for topological indices, it would seem entirely reasonable to believe that their future may be even more promising than their past.

ACKNOWLEGEMENT

The author thanks the organizers of the 1989 Conference of the Chemical Structure Association for making possible the original delivery of this contribution in Durham, England on 19 July 1989.

REFERENCES

[1] R. V. Wagoner and D. W. Goldsmith, *Cosmic Horizons: Understanding the Universe*, Freeman, San Francisco, 1982.

[2] J. Gribbin, *In Search of the Big Bang*, Heinemann, London, 1986.

[3] P. Cloud, *Sci. Amer.* **249**, 176 (1983).

[4] B. J. Carr and M. J. Rees, *Nature* **278**, 605 (1979).

[5] R. H. Dicke, *Rev. Mod. Phys.* **29**, 363 (1957).

[6] J. D. Barrow and F. J. Tipler, *The Anthropic Cosmological Principle*, Oxford University Press, Oxford, 1986.

[7] J. D. Barrow, *The World Within the World*, Clarendon Press, Oxford, 1988.

[8] C. A. Russell, *The History of Valency*, Leicester University Press, Leicester, 1971, chap. 8.

[9] A. M. Butlerov, *Zeitschr. Chem.* **4**, 546 (1861).

[10] A. Crum Brown, *The Theory of Chemical Combination*, M.D. Thesis, University of Edinburgh, 1861; see also A. Crum Brown, *Trans. Roy. Soc. Edin.* **23**, 707 (1864).

[11] G. E. Hein, in *Kekulé Centennial*, Advances in Chemistry Series 61, Am. Chem. Soc., Washington, D.C., 1966, chap. 1.

[12] D. H. Rouvray, in *Annals of Mathematical Chemistry* Vol. 1, Gordon and Breach, London, 1990, chap. 1.

[13] D. H. Rouvray, *Chem. Technol.* **3**, 378 (1973).

[14] G. M. Crippen, *Distance Geometry and Conformational Calculations*, Research Studies Press, Chichester, U.K., 1981.

[15] I. G. Csizmadia, Ed., *Molecular Structure and Conformation*, Elsevier, Amsterdam, 1982.

[16] R. Zahradnik and R. Polak, *Elements of Quantum Chemistry*, Plenum Press, New York, 1980.

[17] J. C. A. Boeyens, *Structure and Bonding* **63**, 65 (1985).

[18] R. E. Merrifield and H. E. Simmons, *Topological Methods in Chemistry*, Wiley, New York, 1989, p. 1.

[19] D. H. Rouvray, *Sci. American* **255**, 40 (1986).

[20] N. L. Biggs, E. K. Lloyd and R. J. Wilson, *Graph Theory: 1736–1936*, Clarendon Press, Oxford, 1976.

[21] R. J. Wilson and L. W. Beineke, Eds., *Applications of Graph Theory*, Academic Press, London, 1979.

[22] A. T. Balaban, Ed., *Chemical Applications of Graph Theory*, Academic Press, London, 1976.
[23] N. Trinajstić, *Chemical Graph Theory*, Vols. I and II, Chemical Rubber Company Press, Boca Raton, Florida, 1983.
[24] L. B. Kier and L. H. Hall, *Molecular Connectivity in Chemistry and Drug Research*, Academic Press, New York, 1976.
[25] L. B. Kier and L. H. Hall, *Molecular Connectivity in Structure–Activity Analysis*, Research Studies Press, Letchworth, England, 1986.
[26] D. H. Rouvray, *J. Mol. Struct. (Theochem)* **185**, 187 (1989).
[27] D. H. Rouvray, *J. Comput. Chem.* **8**, 470 (1987).
[28] M. Randić, *Int. J. Quant. Chem., Quant. Biol. Symp.* **11**, 137 (1984).
[29] J. V. Knop, R. W. Müller, K. Szymanski and N. Trinajstić, *Computer Generation of Certain Classes of Molecules*, SKTH, Zagreb, 1985.
[30] H. Kopp, *Ann. Chem. Pharm.* **50**, 71 (1844).
[31] S. E. Stein and R. L. Brown, in *Molecular Structure and Energetics*, Vol. 2, J. F. Liebman and A. Greenberg, Eds., VCH Press, New York, 1987, chap. 2.
[32] D. Bonchev, O. Mekenyan and N. Trinajstić, *J. Comput. Chem.* **2**, 127 (1981).
[33] J. G. Topliss and R. J. Costello, *J. Med.* **15**, 1066 (1972).
[34] J. G. Topliss and R. P. Edwards, *J. Med. Chem.* **22**, 1238 (1979).
[35] D. Bonchev, *Information Theoretic Indices for Characterization of Chemical Structures*, Research Studies Press, Chichester, England, 1983.
[36] D. H. Rouvray, in *Chemical Applications of Topology and Graph Theory*, R. B. King, Ed., Elsevier, Amsterdam, 1983, p. 159.
[37] S. C. Basak, G. J. Niemi, and G. D. Veith, in *Computational Chemical Graph Theory*, D. H. Rouvray, Ed., Nova Press, New York, 1990, chap. 9.
[38] A. T. Balaban, I. Motoc, D. Bonchev and O. Mekenyan, *Top. Curr. Chem.* **114**, 21 (1983).
[39] D. H. Rouvray, *Congressus Numerantium* **55**, 253 (1986).
[40] P. J. Hansen and P. C. Jurs, *J. Chem. Educ.* **65**, 574 (1988).
[41] M. I. Stankevich, I. V. Stankevich and N. S. Zefirov, *Russ. Chem. Rev.* **57**, 191 (1988).
[42] H. Wiener, *J. Am. Chem. Soc.* **69**, 17 (1947).
[43] H. Hosoya, *Bull. Chem. Soc. Japan* **44**, 2332 (1971).
[44] D. H. Rouvray, in *Mathematics and Computational Concepts in Chemistry*, N. Trinajstić, Ed., Ellis Horwood, Chichester, England, 1986, chap. 25.
[45] J. Plesnik, *J. Graph Theory* **8**, 1 (1984).
[46] R. C. Entringer, D. E. Jackson and D. A. Synder, *Czech. Math. J.* **26**, 283 (1976).
[47] E. R. Canfield, R. W. Robinson and D. H. Rouvray, *J. Comput. Chem.* **6**, 598 (1985).
[48] H. Wiener, *J. Chem. Phys.* **15**, 766 (1947).
[49] J. Walker, *J. Chem. Soc.* **65**, 725 (1894).
[50] D. H. Rouvray and B. C. Crafford, *S. Afr. J. Sci.* **72**, 74 (1976).
[51] N. Trinajstić, [23], p. 116.
[52] D. H. Rouvray and R. B. Pandey, *J. Chem. Phys.* **85**, 2286 (1986).
[53] D. Bonchev, O. Mekenyan, G. Protić and N. Trinajstić, *J. Chromatogr.* **176**, 149 (1979).
[54] M. Medić-Sarić, D. Maysinger and M. Movrin, *Acta Pharm. Jugosl.* **33**, 199 (1983).

[55] S. C. Basak, D. K. Harriss and V. R. Magnuson, *J. Pharm. Sci.* **73**, 429 (1984).
[56] O. Mekenyan, S. Dimitrov and D. Bonchev, *Eur. Polym. J.* **19**, 1185 (1983).
[57] D. Bonchev, O. Mekenyan and H.-G. Fritsche, *Phys. Stat. Sol.* **55A**, 181 (1979).
[58] O. Mekenyan, D. Bonchev and H.-G. Fritsche, *Phys. Stat. Sol.* **56A**, 607 (1979).
[59] O. Mekenyan, D. Bonchev and H.-G. Fritsche, *Zeit. Phys. Chem. Leipzig* **265**, 959 (1984).
[60] D. Bonchev, O. Mekenyan and H.-G. Fritsche, *J. Cryst. Growth* **49**, 90 (1980).
[61] O. Mekenyan, D. Bonchev and N. Trinajstić, *Math. Chem.* **6**, 93 (1979).
[62] D. Bonchev, O. Mekenyan and N. Trinajstić, *Int. J. Quant. Chem.* **17**, 845 (1980).
[63] O. Mekenyan, D. Bonchev and N. Trinajstić, *Int. J. Quant. Chem.* **19**, 929 (1981).
[64] M. Randić, *J. Am. Chem. Soc.* **97**, 6609 (1975).
[65] L. B. Kier, W. J. Murray, M. Randić and L. H. Hall, *J. Pharm. Sci.* **65**, 1226 (1976).
[66] L. B. Kier and L. H. Hall, *J. Pharm. Sci.* **65**, 1806 (1976).
[67] S. C. Grossman, B. Jerman-Blažić and M. Randić, *Int. J. Quant. Chem., Quant. Biol. Symp.* **12**, 123 (1986).
[68] D. H. Rouvray, *Acta Pharm. Jugosl.* **36**, 239 (1986).
[69] R. Kaliszan and H. Lamparczyk, *J. Chromatogr. Sci.* **16**, 246 (1978).
[70] T. DiPaolo, *J. Pharm. Sci.* **67**, 564 (1978).
[71] L. B. Kier, W. J. Murray and L. H. Hall, *J. Med. Chem.* **18**, 1272 (1975).
[72] M. Vighi and D. Calamari, *Chemosphere* **14**, 1925 (1981).
[73] A. Sabljić, *J. Agric. Food Chem.* **32**, 243 (1984).
[74] L. B. Kier and L. H. Hall, *J. Med. Chem.* **20**, 1631 (1977).
[75] S. C. Basak, V. R. Magnuson, G. J. Niemi and R. R. Regal, *Discr. Appl. Math.* **19**, 17 (1988).
[76] I. Motoc and A. T. Balaban, *Rev. Roumaine Chim.* **26**, 593 (1981).
[77] J. T. Edward, *Can. J. Chem.* **60**, 480 (1982).
[78] R. D. Cramer, *J. Am. Chem. Soc.* **102**, 1837 (1980).
[79] R. D. Cramer, *J. Am. Chem. Soc.* **102**, 1849 (1980).
[80] S. El-Basil, in *Graphy Theory and Topology in Chemistry*, R. B. King and D. H. Rouvray, Eds., Elsevier, Amsterdam, 1987, p. 557.
[81] F. Harary and A. J. Schwenk, *Discr. Math.* **6**, 359 (1973).
[82] D. H. Rouvray and S. El-Basil, *J. Mol. Struct. (Theochem)* **165**, 9 (1988).
[83] P. Willett, *Similarity and Clustering in Chemical Information Systems*, Research Studies Press, Letchworth, England, 1986.
[84] D. H. Rouvray, in *Concepts and Applications of Molecular Similarity*, G. M. Maggiora and M. A. Johnson, Eds., Wiley, New York, 1990, chap. 1.
[85] M. S. Lajiness, in *Computational Chemical Graph Theory*, D. H. Rouvray, Ed., Nova Press, New York, 1990, p. 297.
[86] M. A. Johnson, *J. Math. Chem.* **3**, 117 (1989).
[87] D. H. Rouvray and R. B. King, *Am. Chem. Soc. Div. Petrol. Chem. Preprints* **34**, 852 (1989).
[88] M. P. Hanson and D. H. Rouvray, *J. Phys. Chem.* **91**, 2981 (1987).
[89] A. N. Hayhurst and H. R. N. Jones, *J. Chem. Soc. Faraday Trans. II* **83**, 1 (1987).
[90] A. T. Balaban, *Chem. Phys. Lett.* **89**, 399 (1982).
[91] D. H. Rouvray, *J. Chem. Educ.* **52**, 768 (1975).
[92] A. Gräslund and B. Jernström, *Quat. Rev. Biophys.* **22**, 1 (1989).
[93] P. G. Seybold, *Int. J. Quant. Chem., Quant. Biol. Symp.* **10**, 95 (1983).

[94] D. M. Jerina and R. E. Lehr, in *Microsomes and Drug Oxidations*, V. Ullrich, Ed., Pergamon, Oxford, 1977, p. 709.
[95] W. C. Herndon and L. von Szetpály, *J. Mol. Struct. (Theochem)* **148**, 141 (1986).

Similarity-based methods for predicting chemical and biological properties: a brief overview from a statistical perspective

Mark Johnson, Computational Chemistry, The Upjohn Co., Kalamazoo, MI

INTRODUCTION

A number of seemingly diverse methodologies in computational chemistry are being brought together using concepts of molecular similarity [1]. Examples include substructure and similarity searches of chemical databases, cluster and dissimilarity approaches to selecting compounds for screening, property prediction using nearest neighbour and molecular shape analysis methodologies, superpositioning of two chemical graphs or of two 3-D structures, and the prediction of reaction intermediates and generation of synthetic pathways. The purpose of this study is to present similarity-based property prediction as an important example of a molecular similarity space, to suggest a number of emerging statistical methodologies that seem particularly relevant to similarity-based property prediction, and to briefly overview some of the experimental support for the property similarity principle on which such prediction is founded.

I A MOLECULAR SIMILARITY CONTEXT FOR PROPERTY PREDICTION

Although this study concerns similarity-based property prediction methods, it is helpful to locate such prediction methods in the broader contexts of molecular similarity and of chemical property prediction. I will begin by developing the concept of molecular similarity and presenting it as an organizing concept for a number of problems and methodologies in computational chemistry. My approach will largely follow that given in [2].

When someone says two molecules are similar, the natural question is to ask with respect to what. The answer might be with respect to their functional groups, their chemical graphs, their three-dimensional shape, etc. Thus, methods of assessing molecular similarity are at least as varied as the diversity of methods by which we describe molecules. Table 1 presents a few of the many *molecular descriptions* that have been used in the context of molecular similarity analysis. Associated with each molecular description is a mathematical representation that is actually used when computing the similarity between two descriptions. When the molecular description is a vector, the components of the vector are usually referred to as *molecular descriptors*. See Johnson [3] for a more detailed discussion of the mathematical representations and spaces that have been used in molecular similarity analysis.

Table 1 – A list of some chemical descriptions and associated mathematical representations

Chemical description	Mathematical description
Molecular properties	Vectors
Molecular fragments	Vectors
Topological indices	Vectors
Chemical names	Sequences
Constitutional formulas	Graphs
3-D configurations	Finite sets in R^3
Molecular surfaces	Surfaces in R^3
Steric volumes	Scalar fields in R^3
Electron densities	Scalar fields in R^3

Although there is considerable flexibility in selecting a molecular description, there is considerable less flexibility in what similarity concepts are available for comparing the selected descriptions. The first basic concept is that of an equivalence class. An *equivalence class* of molecules consists of those molecules that are necessarily treated in an identical fashion in the subsequent molecular similarity analysis. Each class consists of those molecules given the same molecular description. Thus, if the molecular formula is used as a molecular description, then all isomers of a molecule define the equivalence class containing that molecule. Rouvray [4] gives an extensive list of equivalence classes that have been used in chemistry.

Equivalence classes are possibly the most fundamental concept of similarity, as suggested by their use in defining the important concepts of isomerism. Even so, there is really not much one can say about two molecules using only the concept of equivalence classes based on a single molecular description. One can only say that two molecules either are or are not equivalent with respect to that description. (See Rosen [5] for the rich mathematical theory that can be developed when relating the equivalence classes associated with more than one molecular description.) However, if we take advantage of the molecular description as a rich representation of molecular information, a number of possibilities emerge for relating the associated equivalence classes, namely matchings, partial orderings, and proximity measures.

It is natural to start with the idea of matching two molecular descriptions of the same type since partial orderings and proximity measures are generally defined in terms of matchings. A *matching* of two molecular descriptions can be viewed as a superpositioning of the components of those descriptions. Examples would include the superpositioning of atoms of two atomic configurations so as to minimize the distance between corresponding pairs of atoms, or the matching of the atoms of two chemical graphs so as to preserve as much as possible the bonding arrangements between corresponding atoms.

The important concept of a substructure is defined in terms of matchings. For example, a chemical fragment is a substructure of a molecule if the atoms and bonds of that fragment can all be corresponded to a subset of the atoms and bonds of that molecule. If we view a substructure of a molecule as 'simpler' than that molecule, we set up a *partial ordering* of structures. Partial orders satisfy the transitivity property that if A is 'simpler' than B and B is 'simpler' than C, then A is 'simpler' than C. All substructure searches of a database are founded on the concept of a partial ordering of the structures in that database.

Partial orders also figure into the prediction of reaction intermediates. One can view a sequence of normal alkanes, e.g. methane, ethane, propane, and butane as defining a 'line' of molecules in which one always goes from one molecule to a *next* more complex molecule. This concept of a 'line' of molecules can be generalized by allowing one to go to a *next* simpler molecules also. In this way one can use the partial order to get from any molecule to any other molecule. The Dugundji–Ugi principle of minimal chemical distance essentially states that the intermediates in stoichiometrically balanced reactions usually lie on one of the many shortest 'lines' connecting the reactants and products in such a partial ordering [6].

This principle illustrates the use of a number to indicate the relatedness of two molecules. Following Borg and Lingoes [7], we shall refer to such numbers as *proximity measures*. Proximity measures are broadly divided into two groups: dissimilarity measures and similarity measures. The Dugundji–Ugi distance is an example of a *dissimilarity measure* in which a pair of highly related molecules is assigned a number close to zero. Conversely, a *similarity measure* assigns a pair of highly related molecules a large number, which is usually normalized so that the value is close to 1 as in the case of correlation coefficients. By inverting a similarity measure of relatedness, we obtain a dissimilarity measure. Proximity measures can range from something as simple as a count of the number of atoms and bonds that cannot be matched in the chemical graphs of two molecules to an integral of the difference between two superimposed electron densities. A number of interesting proximity measures can be found in [1, 3, 8].

The relations of matching, partial ordering and proximity together with the function $f{:}M \rightarrow D$ assigning each molecule m in M a chemical description $f(m)$ in D define a simple molecular similarity space apropos the important class of problems associated with the storage and retrieval of structures from a chemical database. Full structure matching, substructure and similarity searches, clustering and dissimilarity selection are some of the interesting similarity operations that are defined on these similarity spaces [8, 9, 10].

If we augment this similarity space with another function $h{:}M \rightarrow R$ which associates a scalar chemical or biological property with each molecule in M, we define the important similarity space of interest to this study, namely a similarity space useful in predicting chemical and biological properties. The value $h(m)$ assigned to molecule m may represent, for example, a melting point, molecular volume or binding coefficient. Interesting examples of chemical descriptions and similarity operations in this context are given in [11, 12, 13]. The next section overviews a number of promising statistical methods relevant to this problem, many of which remain largely unexplored.

A final similarity space that has attracted considerable attention replaces the real line R in the property prediction similarity space with a digraph (V, E) of vertices and arrows.

The function $h:M \rightarrow V$ now maps each molecule in M to a vertex of the digraph. The result is a reaction graph. Each arrow in the reaction digraph indicates a molecule (or reactants) that is transformed into another molecule (or products). This similarity space typifies the important area of computer-aided organic synthesis. The use of lines of compounds defined in terms of partial orders has already been mentioned as an interesting similarity operation relevant to this space. Other uses of matching, partial orders and proximity measures can be found in [6, 14].

II A CLASSIFICATION OF CHEMICAL PROPERTY PREDICTION METHODS

Methods for predicting the chemical and biological properties of a molecule naturally differ with respect to those contexts in which they are particularly appropriate. Similarity-based predictive methods appear to be particularly suitable for the 'automated' association of predicted property values with the compounds in a structure–property/activity database. Such an association can serve a number of useful purposes. First, the number of compounds in the database that have 'useful' property values for a given property can be dramatically extended. Second, the computer can list out structural regions in the database where the property 'surface', to be defined later, has interesting features, for example a steep region or a level region of the surface. Third the computer can flag cases in which there is a disparity between the observed and predicted property value for a molecule, thus indicating either an error in the observed value or a surprising relationship between structure and activity.

For such purposes to be realized, the size and growth rate of structure–property/activity databases requires the development of useful and robust predictive methods largely carried out by the computer with a minimal amount of human intervention. We shall refer to such methods as *automatable*, and state without further justification, that only automatable methods are really suitable for the routine prediction of properties for a large database of compounds.

With this in mind consider a molecular similarity space with a set M of molecules, a set D of corresponding descriptions and a scalar property $h:M \rightarrow R$ that maps each molecule to a real number. In statistical terminology, h is called a *response function*. Fig. 1 illustrates by means of contours a typical response function in which each molecular description is a vector of two descriptors, chemical descriptor 1 and chemical descriptor 2. These two descriptors might themselves be chemical properties of a molecule such as a solubility or molecular volume, or they might represent two *topological indices*, which are simply numbers computed directly for the chemical graph of a molecule, one of the simplest being the number of atoms. Regions A and B denote areas in the molecular similarity space associated with high values for the scalar property.

We can think of the contours of the property values in Fig. 1 as defining a *property surface* over the underlying similarity space. Fig. 1 suggests a 'moderately modal' surface, i.e. a surface with a moderate number of local maxima and minima. At the simpler extreme would be a sparsely modal surface, i.e. a surface with one or two local maxima and or minima or a surface with no local maxima or minima such as a plane surface. At the more complex extreme would be densely modal surfaces. The purpose of this section is to briefly describe a class of property prediction methods appropriate to each degree of mode density: sparse, moderate and dense.

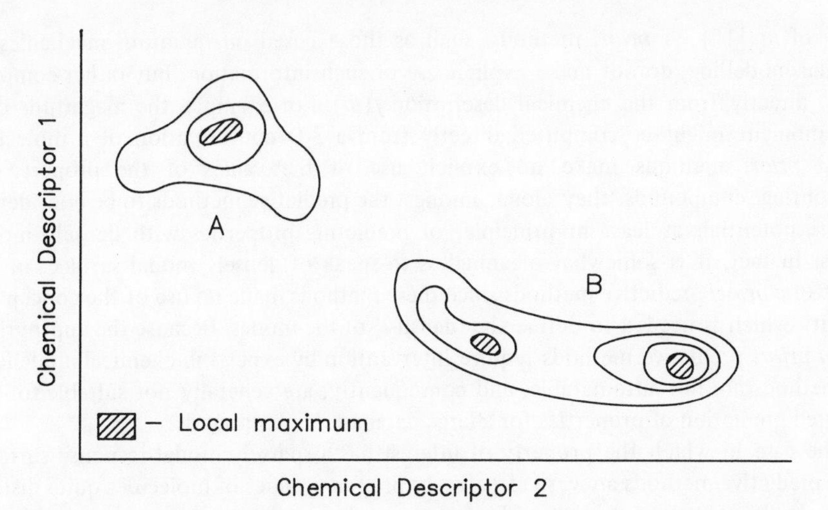

Fig. 1 – Contour description of a response function based on two chemical descriptors.

Fig. 2 illustrates the two components of a function used to predict a scalar property $h(m)$ for molecule m. The first component is the function f that assigns a chemical description $f(m)$ to the molecule m. The second component p maps the description $f(m)$ to the predicted value $p(f(m))$. If $|h(m) - p(f(m))|$ is small for most m, then $p: D \to R$ is a 'good' predictive function.

Methods for predicting the property $h(m)$ have been classified based on the manner in which they take into account the property information available on molecules in the

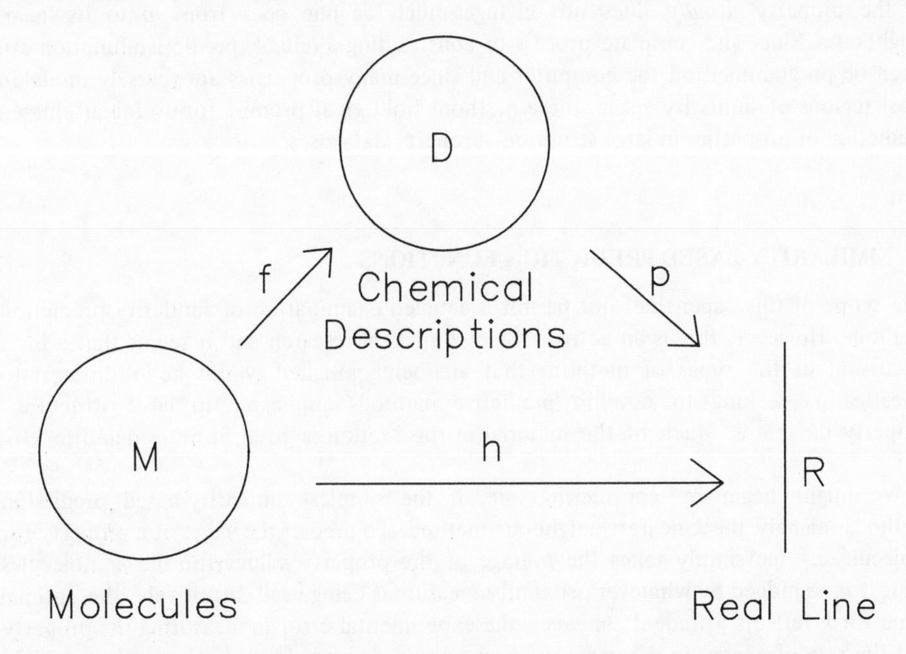

Fig. 2 – Relationship between a property function h, a chemical description assignment function f and a predictive function p.

vicinity of m [15]. *A priori* methods, such as those based on quantum mechanics or molecular modelling, do not make explicit use of such information, but rather compute $p(f(m))$ directly from the chemical description $f(m)$. For example, the magnitude of a dipole moment might be computed directly from a 3-D conformation of a molecule. Since *a priori* methods make no explicit use of the values of the property on 'neighbouring' compounds, they alone, amongst the predictive methods to be considered, have the potential, at least in principle, of predicting properties with densely modal surfaces. In fact, it is somewhat meaningless to speak of densely modal surfaces in the context of *a priori* predictive methods since these methods made no use of the concept of proximity which is needed to define the 'density' of the modes. Because the appropriate use of *a priori* predictive methods require intervention by experts in chemical modelling, these methods are not automatable, and consequently, are generally not suitable for the automated prediction of properties for a large database of compounds.

In the case in which the property of interest has a sparsely modal response surface, 'global' predictive methods are very effective as property values of molecules quite distant from the molecule m of interest can be and are used to predict the property value for m. These methods require that the chemical descriptions be vectors of chemical descriptors and are typified by the standard linear regression models. Global predictive methods are particularly suited for predicting additive properties such as molar refractivity. Because the appropriate use of these predictive methods requires experts in statistical modelling, and because sparsely modal properties are not all that common, global predictive methods are generally not automatable.

Similarity prediction methods lie between these two extremes. Reliable prediction of the value of a property for molecule m requires a knowledge of the value of that property for some molecules 'near' m in molecular similarity space, near enough so that the value of the property *usually* does not change much as one goes from m to its 'near' neighbours. Since the complete process of constructing a reliable predictive function can often be programmed on the computer and since many properties are sparsely modal in most regions of similarity space, these methods hold great promise for useful automated prediction of properties in large structure–property databases.

III SIMILARITY-BASED PREDICTION FUNCTIONS

The scope of this paper does not permit a detailed examination of similarity prediction methods. However, this is an active area of statistical research and it seems that a brief discussion of the types of methods that are being pursued would be of interest to investigators seeking to develop predictive methods applicable to large structure–property databases. Much of the material in this section is given in more detail in [16, 17].

We might begin by considering one of the simplest similarity-based prediction methods, namely the k-nearest neighbour method. To predict the value of a property for molecule m, one simply takes the average of the property values for the k molecules nearest m as judged by whatever proximity measure is being used. Intuitively, the optimal value for k reflects a tradeoff between the experimental error in measuring the property and the rate of change in the property as one proceeds away from m. Generally speaking, the larger the experimental error, the more neighbours should be used so as to reduce the standard error of the predicted value, while on the other hand, the faster the rate of change in the value of the property, the fewer neighbours should be used so as to reduce the bias of the predicted value.

It is helpful to view a k-nearest neighbour prediction of a property value for molecule m as a weighted average of the property values available in the database or dataset. It will simplify the discussion to make a slight change in our notation. We shall write y and d for the observed property value $h(m)$ and the chemical description $f(m)$ of molecule m. Now, suppose we have property values $y_1, \ldots, y_n$ on molecules $m_1, \ldots, m_n$. Then the k-nearest neighbour predictor $p(d)$ for molecule m is given by

$$p(d) = \sum_{i=1}^{n} k^{-1} w_d(d_i) y_i \tag{1}$$

where $w_d(d_i) = 1$ if d_i is one of the k nearest neighbours of d and $w_d(d_i) = 0$, otherwise. The subscript, d, of the weighting function simply reflects the fact that the set of nearest neighbours depends on the molecule m, and consequently, which chemical description d, $d = f(m)$, that is having its associated property value predicted.

The weighting function, $k^{-1}w_d(d_i)$, gives as much weight to the property value of the kth nearest neighbour of d as it does to the property value of the nearest neighbour. On the other hand, the property value of the $(k + 1)$th nearest neighbour is assigned a zero weight. This undesirable situation is remedied by replacing $k^{-1}w_d(d_i)$ by a smoother weighting function which we shall denote by $K(d, d_i)$. Equation 1 now becomes

$$p(d) = \sum_{i=1}^{n} K(d, d_i) y_i \tag{2}$$

where $K(d, d_i)$ is suitably normalized so that

$$\sum_{i=1}^{n} K(d, d_i) = 1.$$

The predictive function in Equation (2) is called a local averaging predictor and $K(d, d_i)$ is called its kernel function. It is natural and customary for $K(d, d_i)$ to attain its maximum when $d_i = d$, i.e. assigns the most weight to the corresponding property value y_i. If $u(d, d_i)$ is a dissimilarity measure, then the inverse of $u(d, d_i)$, after suitable normalization, would be a kernel function having its maximum at $d_i = d$.

Local averaging predictors suffer from what has become known as the 'curse of high dimensionality'. To illustrate, suppose d is an p-dimensional vector $\mathbf{x} = (x_1, \ldots, x_p)$ of chemical descriptors. Suppose the ranges of each of the descriptors is normalized to vary from 0 to 1, and suppose that we have property values associated with 100 molecular descriptions uniformly distributed about that p-dimensional unit hypercube. If we throw down a smaller p-dimensional hypercube, which corresponds to a type of kernel function, whose edges are 0.1 units wide (a tenth of the range of each chemical descriptor), then on average, this smaller hypercube will contain only $(0.1)^p$ of the data points. For example, if $p = 10$, there is only an infinitesimal chance that we would contain any of the 100 data points. On the other hand, if we require that the edges of this smaller hypercube be wide enough so that we generally include 10% of our data points, then the edges must be $(0.1)^{1/p}$ units wide. If $p = 10$, the edges must be roughly 0.8 units wide (80% of the range of each chemical descriptor).

The predictive function arising from the standard regression model can be used to develop a way out of this dilemma, at least in the usual cases when the chemical descriptions are not highly correlated with one another. In our notation, the linear regression function has the form

$$p(\mathbf{x}) = \alpha_0 + \sum_{j=1}^{p} \alpha_j x_j. \tag{3}$$

As pointed out in the preceding section, this predictive function can approximate some sparsely modal response surfaces by letting some of chemical descriptors be quadratic or cubic powers of other chemical descriptors. However, for moderately modal response surfaces, such an approach becomes cumbersome and complex.

By replacing $\alpha_j x_j$ in Equation (3) with a function $g_j(x_j)$, we obtain the more general additive predictive function given by

$$p(\mathbf{x}) = \sum_{j=1}^{p} g_j(x_j). \tag{4}$$

Equation 3 is a special case of Equation 4 in which $g_j(x_j)$ is linearly related to its argument x_j. However, Equation 4 allows each $g_j(x_j)$ to be a complex multimodal function of its argument and lets these functions vary considerably from one chemical descriptor to the next. This predictive function can be viewed as a similarity-based predictive function because the form of each g_j in a neighbourhood of x_j is determined by the property values associated with those chemical descriptions which are similar with respect to their jth component.

Even with this flexibility, there are many moderately modal response surfaces that the predictive function given by Equation 4 cannot approximate well. Fig. 1 is just one example. To see this, consider what the response surface of an additive function of chemical descriptors 1 and 2 must look like. For example, suppose we look at two vertical slices through response surface defined by two fixed values for chemical descriptor 2. They must have exactly the same shape since they can differ only by a constant defined by the difference in the value of g_j at the two fixed values. Since the vertical slices in Fig. 2 differ in shape, the response surface in that figure cannot be modelled by an additive predictive function. The preceding argument for two chemical descriptions can be stated relative to p dimensions. The 'shapes' of $p(\mathbf{x})$ in $p-1$ dimensions for two fixed values x_j and x_j' of the jth chemical descriptors must be identical up to a constant given by $g_j(x_j) - g_j(x_j')$.

Significantly more general predictive functions can be defined by replacing the chemical descriptors $x_1, \ldots, x_p$ with a new set of chemical descriptors $z_1, \ldots, z_q$ where each z_j is a linear combination of $x_1, \ldots, x_p$. The particular linear combination depends on the shape of the response surface. This new predictive function is given by

$$p(\mathbf{x}) = \sum_{k=1}^{q} g_k(z_k) \tag{5}$$

where

$$z_k = \sum_{j=1}^{p} \alpha_{jk} x_j.$$

Unlike principal components, these new 'axes' need not be orthogonal and their number, q, may exceed p. The predictive function in Equation 5 is called a projection pursuit model.

Friedman [16] discusses a number of pros and cons of each of these models. In particular, he notes that any reasonably smooth response surface can be approximated by a predictive function of the form given in Equation 5. On the other hand, some rather simple response surfaces can require quite a large value for q for a good approximation.

Friedman's own research involves a generalization to higher dimensions of a spline approach to smoothing functions in one dimension. It is beyond the scope of this article to present this and a number of other methodologies that are emerging in this active area of statistical research. These few examples have been presented to give some of the flavour of these similarity-based approaches that are becoming available.

Before leaving this section, it might be noted that the additive and projection pursuit models assume the mathematical representation of the chemical descriptions be vectors. As noted in Table 1, this is certainly not always the case. However, if one has a proximity measure, multidimensional scaling techniques [18, 19] can be used to map the chemical descriptions to an n-dimensional Euclidean space so as to maximally preserve the proximity of the various pairs of molecules, at least for small datasets. The preceding vector-based predictive methods then become available for property prediction.

IV THE SIMILAR PROPERTY PRINCIPLE

None of these predictive methods will give reliable predictions if the similar property principle is not satisfied [20]. This principle states that similar structures generally have similar properties. Of course, the validity of this principle is likely to depend on the particular combination of chemical descriptions, proximity measures, molecules and properties under consideration. This section briefly overviews some of the initial efforts in establishing the validity of this principle and, conversely, of using the similar property principle as a basis for evaluating the utility of various chemical descriptions and proximity measures.

The first general study of the similar property principle involving a broad mix of descriptions, proximity measures and properties was carried out by Willett and Winterman [21]. In this study a diverse mix of 16 different properties were considered including p*I* values on 20 naturally occurring amino acids, heats of vaporization for 129 alkanes, alkenes, alcohols, ketones and benzene and pyridine derivatives, inhibition of chlorphentermine binding in rat lung by 20 structurally diverse compounds, and toxicity to mice of 25 aliphatic and carbocyclic ethers. In addition, 12 different vector-based chemical descriptions were examined which involved two different types of chemical descriptors, the augmented atom fragment and the atom–bond–atom fragment, each examined with respect to six different weighting schemes. Six different correlation and distance coefficients defined on vectors were then tried in combination with each of the 12 possible chemical descriptions giving rise to 72 different proximity measures.

Predicted values were obtained for each molecule in each of the 16 datasets using nearest neighbour prediction after disallowing a molecule from being its own nearest neighbour. In this way, a predicted value could be obtained for each molecule in the dataset. The correlation between the observed and predicted property values was then determined for each proximity measure. The reader is referred to the original article for a detailed discussion of the results. Here we simply note that statistically significant correlation coefficients were obtained for each of the datasets and in many cases these correlation coefficients exceeded 0.8.

More importantly, the authors were able to draw some general conclusions concerning the relative merit of some of the proximity measures. In particular it was found that proximity measures based on correlation coefficients generally outperformed those based on distance coefficients. It was also found that a simple listing of the occurrence frequency of each fragment worked as well as or better than a number of the more complex weighting schemes. Johnson *et al.* [22] extended this study for eight of the smaller datasets to include a non-vector-based proximity measure which involves computing the maximum common substructure of two molecules. In this case the correlation coefficients ranged from 0.6 to 0.9. This proximity measure performed consistently better than the best of the preceding 72 fragment-based proximity measures. Although the substructure proximity measure can take orders of magnitude longer to compute than a fragment-based measure, the latter result illustrates how strongly the validity of the similarity principle can depend on the choice of a chemical description and proximity measure.

In a related study of clustering methods involving essentially the same collection of datasets, Willett *et al.* [23] replaced the nearest neighbour predictive function with the average of the property values for those compounds occurring in the same cluster as the compound whose property was being predicted. As we noted earlier, including more neighbours in the predictive function lowers the variance of the predictor at the risk of increasing its bias. With those particular datasets, the tradeoff favoured the reduction of variance so that considerably better correlation coefficients were obtained. However, the authors noted in a later conversation that the tradeoff need not favour the reduction of variance if the clusters get too large as can often occur for large datasets. Thus we see that the validity of the similarity principle can also dramatically depend on the choice of a predictive function, as the preceding section would suggest.

Whereas these studies involve properties that can take on a wide range of values, other cases exist in which the available data indicates only the activity or inactivity of a compound. The question then arises as to whether or not the 'actives' occur together in subregions of the description space or are randomly dispersed about that space. In the first case, cluster- and dissimilarity-based methods of selecting compounds for screening can be expected to turn up a higher rate of novel leads than traditional random selection methods [23, 24]. Without going into the details in this article, Johnson and Lajiness [25] showed that indeed actives do cluster in chemical description space and the degree of clustering can depend on the proximity measure employed.

Clearly, more investigations will be needed to empirically and objectively validate the similarity principle and to select the most appropriate proximity measures and similarity-based prediction function. However, these early studies hold out the promise that automated property prediction based on similarity methods will be sufficiently reliable to find a number of useful applications in research institutions where large structure–property databases exit.

REFERENCES

[1] Johnson, M. A. and Maggiora, G. M., eds., 1990, *Concepts and Applications of Molecular Similarity Analysis*, Wiley Interscience, New York.

[2] Maggiora, G. M. and Johnson, M. A., 1990, Chap. 1 in [1].

[3] Johnson, M. A., 1989, *J. Math. Chem.*, **3**, 117–145.

[4] Rouvray, D. H., 1990, Chapter 2 in [1].

[5] Rosen, R., 1990, Chap. 12 in [1].

[6] Ugi, I., Wochner, M., Fontain, E., Bauer, J., Gruber, B. and Karl, R., 1990, Chap. 9 in [1].

[7] Borg, I. and Lingoes, J., 1987, *Multidimensional Similarity Structure Analysis*, Springer-Verlag, New York, p. 4.

[8] Willett, P., 1987, *Similarity and Clustering in Chemical Information Systems*, Research Studies Press Ltd., Letchworth, 254pp.

[9] Willett, P., 1990, Chap. 3 in [1].

[10] Bawden, D., 1990, Chap. 4 in [1].

[11] Randić, M., 1990, Chap. 5 in [1].

[12] Hopfinger, A. J. and Burke, B. J., 1990, Chap. 7 in [1].

[13] Carbó, R. and Calabuig, B., 1990, Chap. 6 in [1].

[14] Johnson, M. A., Gifford, E. and Tsai, C.-c., 1990, Chap. 10 in [1].

[15] Johnson, M., Basak, S. and Maggiora, G., 1988, *Math. Comput. Modelling,* **11**, 630–634.

[16] Friedman, J. H., 1988, *Fitting Functions to Noisy Data in High Dimensions*, SLAC-PUB-4676, Stanford, 1–35.

[17] Friedman, J. H. and Silverman, B. W., 1989, *Technometrics,* **31**, 3–21.

[18] Davison, M. L., 1982, *Multidimensional Scaling*, John Wiley & Sons, New York, 242pp.

[19] Kruskal, J. B. and Wish, M., 1978, *Multidimensional Scaling*, Sage Publications, Beverly Hills, 93pp.

[20] Wilkins, C. L. and Randić, M., 1980, *Theoret. Chim. Acta,* **58**, 45–68.

[21] Willett, P. and Winterman, V., 1986, *Quant. Struct.–Act. Relat.*, **5**, 18–25.

[22] Johnson, M. A., Naim, M., Nicholson, V. and Tsai, C.-c., 1987, Ed. D. Hadži and B. Jerman-Blažić, In *QSAR in Drug Design and Toxicology*, Elsevier Science Pub., Amsterdam, 67–69.

[23] Willett, P., Winterman, V. and Bawden, D., 1986, *J. Chem. Inf. Comput. Sci.,* **26**, 109–118.

[24] Lajiness, M. S., Johnson, M. A. and Maggiora, G. M., 1989, Ed. J. L. Fauchère, *QSAR: Quantitative Structure–Activity Relationships in Drug Design*, Alan R. Liss, Inc., New York, pp. 173–176.

[25] Johnson, M., Lajiness, M. and Maggiora, G., 1989, Ed. J. L. Fauchère, *QSAR: Quantitative Structure–Activity Relationships in Drug Design*, Alan R. Liss, Inc., New York, pp. 167–171.

Integrated systems

Progress towards integrated chemical information systems

David Bawden, Pfizer Central Research, Sandwich, Kent, England.

FORMS OF INTEGRATION

The somewhat over-used term 'integration' is used to describe many types of chemical information system. Indeed, it has become almost an obligatory piece of sales hype for any chemical information system worthy of being brought to market. This particularly so, given the great interest in the integration of chemical information systems with other computer systems, which has been seen since the mid-1980s.

Several attempts have been made to define and categorize integration, in the specific context of chemical information systems. Williams and Franklin [1], for example, think of three axes, along which integration may be said to occur: connectivity, coherence, and cognition. Hagadone [2] defines five levels of integration, in increasing order of convenience for the user:

no integration
all data in structural database
integration through 'customer supervisor' program
common interface for all data through structure searching system
structurally extended relational database management systems (DBMSs)

In this presentation, I shall use a ranking of four types of integration, in increasing order of 'tightness' or 'fullness' of integration, as it would appear to a user:

ASCII file interchange
common file format
direct interaction
single apparent system

In general, the further down this list, which is actually a continuum, we go, the more advantages to the user are given by integration. These are seen in terms both of speed and convenience of operation, and of additional facilities conferred. The first two correspond roughly to Williams and Franklin's 'connectivity', the latter two to their 'coherence'. Aspects of integration, particularly at the connectivity levels, are discussed in [3].

ASCII file interchange

This is the 'lowest' common denominator form of integration. Data produced by one program or system can be written out to an intermediate file, to be read in by another. It is a simple procedure, and confers a degree of flexibility, since there needs to be no real interaction between the different software systems. It is, however, fiddly, inconvenient, time-consuming, error-prone, and unattractive to most chemist end-users. It will generally either require some modifications to be made to the systems used, or will require the users to work at the operating system level, renaming files and so on. There will also almost certainly be problems of inconsistencies in file formats between different systems.

Common file format

This overcomes the latter problem of ASCII file integration, by providing a consistent format for describing molecular structures and associated data, common to a number of software systems, and ideally to all that are being used in a given environment.

This solution should be available whenever several software systems are purchased from a single supplier, even if a more elaborate form of integration is available. Where systems from several suppliers are in use, there has to be some agreed standard for information exchange. A number of *de facto* standards have emerged in the chemical information area, but have now been superseded by the Standard Molecular Data (SMD) format [4].

Even with an agreed standard such as SMD, integration by file exchange still limits the value added by integration, for the reasons noted in the previous section.

Direct interation

A next stage in completeness of integration is the direct interaction between software systems. One system may, for example, be called from within another, or an 'umbrella' system may give access to a number of specialized routines. In either case, information and file transfer is carried out in a way transparent to the user.

Nonetheless, the user remains clearly aware which system is being used at any time, and needs to be familiar with the interfaces and languages of all the systems which are to be used.

Single apparent system

The fullest form of integration occurs when the user is not necessarily aware of which software system is being used. Not only is information being transferred automatically and transparently, but all systems are accessible by a common interface.

There may be some dispute as to whether this degree of integration is actually desirable in all cases, since it may lead to a lack of consideration of the appropriateness of the methods and routines being applied. This may be especially so, when techniques for modelling and property prediction are being used, more than simply routines for retrieval and display. It may be positively dangerous to provide simple means for estimating numeric values, without also providing the user with an assessment of the validity of the result. This implies some information on the strengths, weaknesses, and limits of applicability of the methods being used.

This problem is likely to be exacerbated as consistent interfaces and tight integration bring sophisticated routines to users who are not specialists in those areas. Explanatory and tutorial guidance, far beyond conventional help facilities, will be a vital part of systems with this quality of integration. They may well need to incorporate some degree of artificial intelligence, bringing in Williams and Franklin's idea of the 'cognitive' dimension of integration [1]. An elaborate data dictionary framework will also be essential, as more systems and data types are integrated.

This is not at all to take the Luddite point of view that integration is a bad thing, since difficulty in using systems restricts their availability to the 'safe' circle of experts and specialists. The point is simply that additional user support facilities must accompany integration, if systems are to be used effectively.

In past years, the difficulty of using chemical information systems based upon arcane codes and notations, and involving intractable batch processing systems, ensured that these systems were used only by information professionals, and a few hardy users. Graphical interfaces and interactive searching procedures have opened these systems up to a much wider user community, with generally agreed benefits. The emerging generations of fully integrated chemical information systems have the potential to provide a similar broadening of the use of all the techniques of computational chemistry.

LEVELS OF INTEGRATION

In speaking of integration within chemical information systems, we are in practice speaking of the integration of various types of information and data around the common core of the chemical structure diagram. Before considering the ways and means of such integration, it should be noted that this integration can occur at various levels *within* the structure diagram representation. This is to say, information may relate to the whole chemical entity, as represented by its structure diagram, or to some substructural element, or conversely to a superstructural class of compounds.

We may distinguish five levels of integration, beginning with the smallest substructural element:

atom/bond
group/substructure
molecular
reaction/synthesis
class/group

Atom/bond

This is the lowest level at which chemical information can reasonably be said to reside. Atom and bond properties are generally numerical, e.g. partial charge, electronegativity, co-ordinate position. Atoms and bonds may also carry simple text labels, e.g. atomic number, bond type, stereochemistry indicator.

Group/substructure

Data relating to larger substructures will also be mostly numeric, e.g. group contributions to additive properties such as molar refractivity or partition coefficient, or spectral data. Such information may also be qualitative and textual, e.g. synthon, bioisostere, or descriptive or physicochemical property; bulky, lipophilic, etc.

Molecular

At this level are those properties assigned to a specific chemical substance. This is, of course, the richest level of integrated information, including numeric, coded, and textual information.

Reaction/synthesis

Information relating to the ensemble of several structures in a reaction scheme, or synthetic sequence, is largely numeric, e.g. yield, reaction conditions, but also includes qualitative and coded information, e.g. stereospecificity, hazard potential, literature reference, experimental procedure, ease and elegance of synthetic plan.

Class/group

At this broadest level, associated information is likely to be largely qualitative and textual in nature: activities and uses, natural occurrence and synthetic routes, etc. Numeric data will largely and necessarily be expressed as ranges and distributions.

So it is clear that information and data must be integrated at a variety of different levels of structural description. Something of a trend can be seen, in that moving from the substructural to the superstructural is likely to parallel a transition from the quantitative and specific to the qualitative, indefinite and textual. This is in accordance with the observation that systems for handling submolecular information are largely of the DBMS kind, e.g. spectral databases [5], whereas information on groups and classes of compounds is to be found mainly in the text handling systems of 'conventional' publishing.

For information and data tied to the individual molecule, which includes most non-structural information in most chemical information systems, it is clear that both text retrieval and DBMS functions are needed to provide a fully adequate set of retrieval functions. Perhaps the best illustrations of this are the Beilstein database, which includes a variety of structurally related data types in DBMS format [6], and the TOSAR system of concept specification [7]. True text retrieval linked to chemical structure information is not as yet available.

The same is true for reaction/synthesis information, though the information systems here are not yet as large or as diverse as for single structures. It is worth noting that the developers of the ORAC reaction retrieval system found it worth developing an elaborate subject description language, to handle qualitative concepts of reaction information which could not be well expressed graphically [8].

This trend, going from quantitative to qualitative as the level of integration changes from substructure through structure to superstructure, is far from being an absolute distinction or progression. We must conclude that the future of integrated systems will require an ability to relate any form of information and data to any level of molecular description.

INFORMATION TO BE INTEGRATED

We should now ask, in concrete terms, just what types of information and data are likely to require integration. Without attempting a complete listing, we can give the more significant classes as:

2-D structure

3-D conformation

generic structure

reaction/synthesis

semi-structural concepts

text (descriptor, keyword, reference)

numeric

graphic images

The first class are descriptions of chemical structure themselves.

2-D structure

3-D conformation

reaction/synthesis

generic structure

The first, the 2-D structure diagram, comprises the central, or core, representation, to which the remainder are linked. 3-D structure may be represented in many ways, static or dynamic, and may reflect a single conformation, a variety of feasible conformations, or some flexible model. The third and fourth classes are superstructural ensembles, either of structures *per se* or of an inter-related reaction sequence. These may be stored either as some form of set of individual representations, with linking information, or as a single generic (Markush) representation.

The next class, *semi-structural concepts,* relates to such things as qualitative descriptions of molecular structure: electron-releasing group, bulky substituent, flexible chain, labile group, protecting group, etc. They will generally be a part of a generic structural description, or associated information for a 2-D structure.

There remain three general forms of information, which, as we have seen, can relate to structural information at varying levels of specificity:

Text

Numeric/coded

Graphic images

Text may be structure element (name, supplier, descriptor, keyword, reference, etc.) applying to a single structural entity. Alternatively it may be genuinely unstructured text, in which links to structural entities are effectively embedded.

Numeric and *coded* information will generally have a straightforward one-to-one correspondence with structural entities, though as we have said this may be at one or more of a number of possible levels of structural specificity.

Finally, *graphic images* (structure diagrams, reaction schemes, analytical traces, full-text, etc.) may similarly relate to chemical entities at any of the possible levels.

INTEGRATION OF SYSTEM FUNCTIONS

It is clear that few, if any, chemical information systems have made very much progress towards the goal of full integration at all levels of the kinds of information and data noted above. It is worthwhile thinking of the kind of integration of system functionality which would be needed to achieve this, though not in any degree of detail.

A crude diagrammatic representation is shown in Fig. 1. This shows a database of chemical structures acting as the core of the system, linking to the systems which surround it, and acting as the interface and gateway between them.

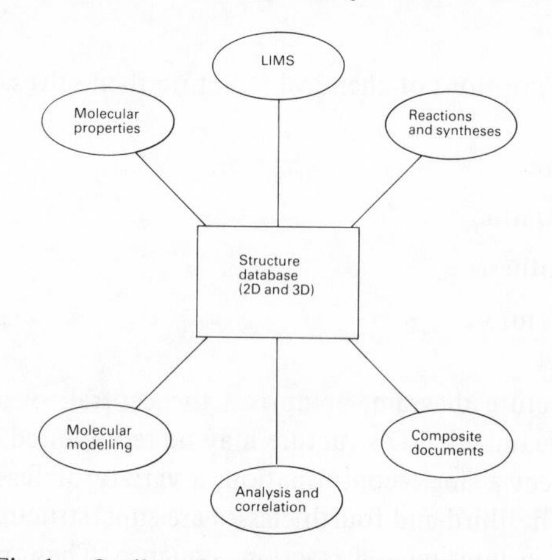

Fig. 1 – Outline of integrated system components.

In most cases, the database will be a full structure storage and retrieval system, since this will satisfy most needs for the majority of chemical information systems. In some cases, e.g. patent systems, the central core will be a generic structure storage and retrieval system. This central system will incorporate both 2-D (structure diagram) and 3-D (conformation) searching, with automatic generation of coordinates from the structure diagram representation, when experimental values are not available. The 2-D representation thus forms the basis for the whole combination of systems, the remainder of its activities indeed being 'beyond the structure diagram', though crucially dependent on it.

Linked to this core system and, via it, to each other will be a number of other systems. Their nature will differ according to the needs of the organisation, but the point is exemplified by a set of six linked functionalities, such as might be appropriate for a pharmaceutical or fine chemical research organization.

Starting at the top, and proceeding anticlockwise, the first linked system is a *laboratory information management system* (LIMS). A number of these products are now entering widespread use, though there is little agreement as to what would correspond to a full and formal definition of a LIMS, each system at present having its own strengths and weaknesses. For this illustration, I take LIMS to imply a system which, among its other functions, holds all the raw and summarized data from in-house investigation, together with details of procedures and protocols. This may be regarded as the long awaited 'electronic laboratory notebook', together with the necessary features for validation of results. In the context of the overall information system, the LIMS component, linked to the core structure retrieval system, makes all in-house generated compound-related information available. It will also provide access to computer-aided structure elucidation techniques.

The second component is summarized as *molecular properties*. This describes a system, or more probably a set of systems, for storage, retrieval, and calculation of quantitative and qualitative properties, and any levels of molecular specificity. The retrieval components will have to include both DBMS and text retrieval functionality. With both measured and predicted values available from such a system, questions of data validation and quality assurance will assume importance.

The *molecular modelling* function overlaps with this to some extent, since this also calculates molecular properties from structure. It is distinguished here, since it primarily refers to routines for conformation analysis, molecular mechanics and dynamics, etc. which would be used for examining conformation factors, with molecular graphics displays.

These latter two functions, molecular property estimation and retrieval and molecular modelling, overlap with the next, *analysis and correlation*. This describes the use of statistical and classificatory routines to examine relationships between structure and property.

The next component, *composite documents*, suggests the role of the structure database in linking together information which will be combined with other material to form reports, publications, etc. This will require a combination of 'live' and 'dead' links to material in other databases, but it is probable that the structure diagram database will retain central importance in any system primarily dealing with chemical documents of whatever sort. Initial approaches to this difficult problem are discussed in a series of conference papers [9].

Finally, the *reactions and syntheses* component includes those systems which deal with the inter-relation of structures within a synthetic ensemble: reaction retrieval programs, synthetic analysis systems, reaction prediction routines, etc.

In this schematic overview of a fully integrated system, no attempt has been made to define precisely what functionality each component would possess, still less to describe their operations in detail. Two points should, nonetheless, be made.

First, the closer that the linkages between the components can come to the level of the '*single apparent system*', the more effective will such a system be, and the more acceptable to its users.

Second, each of the systems must offer, for maximum utility, a variety of levels and modes of access, to suit different types of user. In the case of retrieval systems, for example, this means permitting accessing by different forms of search, including novice and expert modes, browsing, and 'query by example'.

APPROACHES TO INTEGRATED SYSTEMS AT PFIZER CENTRAL RESEARCH

By way of illustration of some of the points made in generality above, progress towards integrated chemical information provision at Pfizer Central Research, Sandwich, will now be mentioned.

The core of all chemically related information systems is SOCRATES, the Pfizer chemical and biological databank system [10]. Integration has been a major feature of the design of SOCRATES since its inception. This has meant in particular:

integration of chemical structure information with biological and physical data;

ability to integrate external routines and procedures into the SOCRATES environment;

presentation of a single user interface, so far as this is feasible.

Technically, the means to accomplish this have included adoption of a single DBMS for all forms of data, an inverted bit-map search engine, and structural access both by conventional substructure search and by browsing.

The linking of other types of system into SOCRATES is shown in outline in Fig. 2.

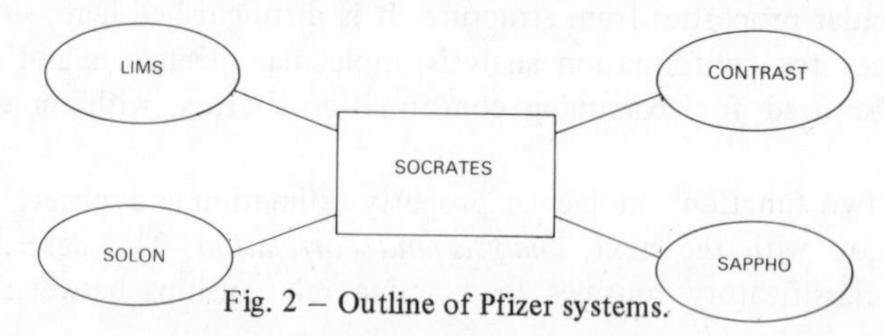

Fig. 2 – Outline of Pfizer systems.

The most significant functionalities interacting with SOCRATES include:

LIMS	a variety of laboratory data systems.
CONTRAST	a SOCRATES-like reaction retrieval system. Further links into other systems for computer-aided synthetic planning are under development.
SOLON	a 3-D conformational searching system.
SAPPHO	an umbrella system for the integration of a variety of molecular modelling and correlation/analysis routines with database functions.

Linking of structural information into text documents is also under investigation, as are gateway systems to literature databases.

CONCLUSIONS

Integrated chemical information systems going 'beyond the structure diagram' must incorporate many diverse kinds of information and data, at a variety of levels of structural specificity. Integration will only be truly effective when it operates at the level of the 'single apparent system', with a high degree of what Williams and Franklin categorize as 'coherence', and elements of their 'cognitive' integration. There is a strong need for elaborate guidance/tutorial/explanatory/advisory functions, and for a comprehensive data dictionary.

This sort of integrated system will include two distinct, though overlapping, elements. First, there will be advanced retrieval functions, beyond substructure search. These will include:

- conformation searching
- reaction retrieval
- integrated structure/data retrieval
- bioisosteric retrieval
- analogy finding by many routes
- gateways to other retrieval systems, especially literature databases

Second, there will be linked calculation, prediction and correlation tools, such as:

- molecular orbital calculations
- molecular mechanics and molecular dynamics
- substructural analysis and topological indices
- multivariate statistics and chemometrics
- toxicity and hazard prediction
- synthetic analysis
- structural elucidation

These two types of functionality will be linked to the 'core' structure handling procedures, by a series of 'intuitive' user interfaces, by advanced graphical displays, and by transparent data transfer.

Advances in technology, and especially the move towards open systems with associated standards for information representation and communication, will play a major part in bringing this about. Even more crucial, however, is the form of knowledge representation and structuring. It is here that the centrality of routines for handling the structure diagram representation of chemical structure becomes evident. This notation, more than a century old, has retained its value in two quite distinct ways. For communication of knowledge, it remains the single most important means of conveying chemical information, an importance which has only been emphasized by the development of graphical structure handling systems. For the expression for a physical reality, the structure diagram encodes a wealth of information, both explicitly and implicitly, as its continuing, indeed expanding, use in correlation and analysis testifies. In both these aspects, use of the structure diagram representation leads, both conceptually and computationally, to more sophisticated structural descriptions. It will remain, therefore, the basis for advanced chemical information systems of the future.

REFERENCES

[1] M. Williams and G. Franklin, 'Future directions in integrated information systems: is there a strategic advantage?', in *Chemical structures. The international language of chemistry*, W. A. Warr (ed.), Berlin, Springer Verlag, 1988.

[2] T. R. Hagadone, 'Current approaches and new directions in the management of in-house chemical structure databases', in *Chemical structures. The international language of chemistry*, W. A. Warr (ed.), Berlin, Springer Verlag, 1988.

[3] W. A. Warr (ed.), *Chemical structure information systems. Interfaces, communication, and standards*, ACS Symposium Series No. 400, Washington, American Chemical Society, 1989.

[4] J. Barnard, 'Draft specifications for the revised version of the Standard Molecular Data (SMD) format', *Journal of Chemical Information and Computer Sciences*, 1990, **30**, 81–96.

[5] W. Bremser, 'Substructural analysis as basis for intelligent interpretation of spectra', in *Chemical structures. The international language of chemistry*, W. A. Warr (ed.), Berlin, Springer Verlag, 1988.

[6] C. Jochum, 'Building a structure-oriented factual databank', in *Chemical structures. The international language of chemistry*. W. A. Warr (ed.), Berlin, Springer Verlag, 1988.

[7] R. Fugmann, 'Grammar in chemical indexing languages, in *Chemical structures. The international language of chemistry*, W. A. Warr (ed.), Berlin, Springer Verlag, 1988.

[8] A. P. Johnson and A. P. Cook, 'Automatic keyword generation for reaction searching', in *Modern Approaches to Chemical Reaction Searching*, P. Willett (ed.), Aldershot, Gower, 1986.

[9] W. A. Warr (ed.), *Graphics for chemical structures: integration with text and data*, ACS Symposium Series No. 341, Washington, American Chemical Society, 1987.

[10] D. Bawden, T. K. Devon, D. T. Faulkner, J. D. Fisher, J. M. Leach, R. J. Reeves and F. E. Woodward, 'Development of the Pfizer integrated research data system W. A. Warr (ed.), Berlin, Springer Verlag, 1988.

Index

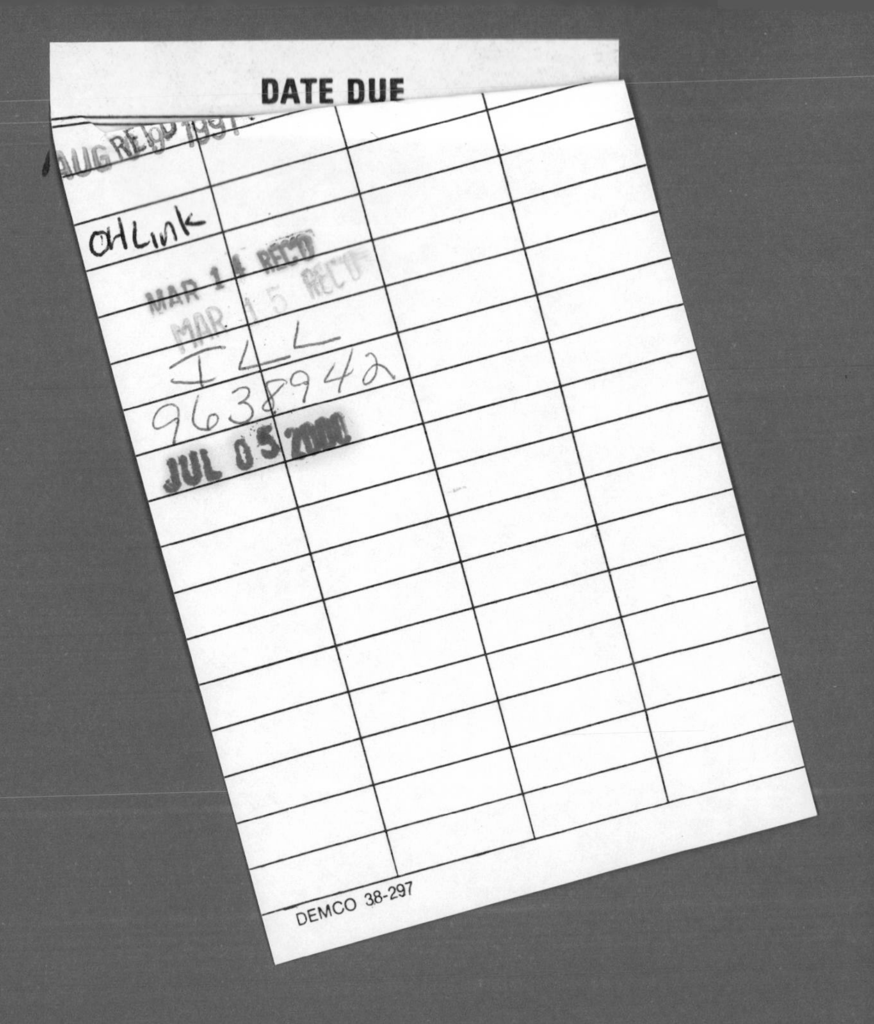
DATE DUE
OhioLINK
ILL
9638942
JUL 05 2000
DEMCO 38-297